Plant Gene Research

Basic Knowledge and Application

Edited by

E.S. Dennis, Canberra
B. Hohn, Basel
Th. Hohn, Basel (Managing Editor)
P.J. King, Basel
J. Schell, Köln
D.P.S. Verma, Columbus

Springer-Verlag Wien NewYork

Molecular Approaches to Crop Improvement

Edited by
E.S. Dennis and D.J. Llewellyn

Springer-Verlag Wien New York

Dr. Elizabeth S. Dennis
Dr. Danny J. Llewellyn

CSIRO Division of Plant Industry
Canberra, ACT

With 34 Figures

Softcover reprint of the hardcover 1st edition 1991
Typeset by Macmillan India Ltd., Bangalore 25

Library of Congress Cataloging-in-Publication Data
Molecular approaches to crop improvement/
edited by E.S. Dennis and D.J. Llewellyn.
p. cm. — (Plant gene research)
Includes bibliographical references and index.
ISBN-13: 978-3-7091-9110-1
1. Crops — Genetic engineering. I. Dennis, E.S. (Elizabeth S.), 1943– . II. Llewellyn, D.J. (Danny J.), 1954– . III. Series.
SB123.57.M65 1991 90-24701
631.5′23 — dc20 CIP

ISSN 0175 2073
ISBN-13: 978-3-7091-9110-1 e-ISBN-13: 978-3-7091-9108-8
DOI: 10.1007/978-3-7091-9108-8

Preface

Although plant genes were first isolated only some twelve years ago and transfer of foreign DNA into tobacco cells first demonstrated some eight years ago, the application and extension of biotechnology to agricultural problems has already led to the field-testing of genetically modified crop plants. The promise of tailor-made plants containing resistance to pests or diseases as well as many other desirable characteristics has led to the almost compulsory incorporation of molecular biology into the research programs of chemical and seed companies as well as Governmental agricultural agencies.

With the routine transformation of rice and the early evidence of transformation of maize the possibility of the world's major cereal crops being modified for improved nutritional value or resistance characteristics is now likely in the next few years.

The increasing number of cloned plant genes and the increasing sophistication of our knowledge of the major developmental and biochemical pathways in plants should eventually allow us to engineer crop plants with higher yields and with less detrimental impact on the environment than now occurs in our current high input agricultural systems. This book draws together many of the expanding areas of plant molecular biology and genetic engineering that will make a substantial contribution to the development of the more productive and efficient crop plants that the world's farmers will be planting in the next decade.

Canberra, December 1990 E.S. Dennis and D.J. Llewellyn

Contents

Chapter 1

Transgenic Rice Plants

Ko Shimamoto

Plantech Research Institute, 1000 Kamoshida, Midori-ku, Yokohama, 227 Japan

With 5 Figures

Contents

I. Introduction

Transgenic plants are becoming indispensable tools for studies of gene regulation in higher plants (Schell, 1987; Benfey and Chua, 1989). Furthermore, with the rapid progress in isolation and characterization of plant genes, it is considered to be one of the most promising approaches for crop improvement (Gasser and Fraley, 1989). Until recently, however, most of the studies using transgenic plants employed Solanaceae species such as tobacco and petunia because of the ease of their transformation. In dicotyledonous species, it has been observed that some monocotyledonous genes, derived from the Graminaceous species, are not expressed at all (Keith and Chua, 1986; Ellis et al., 1987) or when expressed, their expression is not properly regulated. Although the addition of enhancer sequences can sometimes stimulate expression of monocot genes (Ellis et al., 1987) in transgenic tobacco, a model monocot species, in which monocot gene expression can be routinely analysed, is needed.

For cereal crop improvement, the application of genetic engineering techniques developed in model dicot species has so far been precluded until recently, because of the lack of a reliable gene transfer system in cereals. Rice is one of the most important crops in the world and its genome is one of the smallest among important crop species (Bennett et al., 1982). Furthermore, recent progress in genetic manipulation with protoplasts makes rice a suitable candidate in which expression of monocot (cereal) genes can be studied.

In this paper I will first briefly describe protoplast culture of rice, because this is the basis for the progress in the routine production of transgenic plants. Then, techniques of gene transfer applied to rice and other major cereals will be examined, and two examples of foreign gene expression in transgenic rice plants presented. These examples suggest that rice is now a useful system for the expression of monocot genes.

II. Rice Protoplast Culture

Rice protoplasts competent in both division and plant regeneration can be isolated in large quantity from "embryogenic" suspension cultures derived from mature seeds. Isolated protoplasts are able to divide in 3–4 days and form colonies with an efficiency of 1–10% depending on the genotype, quality of the suspension and other factors (Kyozuka et al., 1987, 1988). The presence of nurse cells is beneficial and essential in some cultivars to obtain reproducible plating efficiencies (Kyozuka et al., 1987). This effect of nurse cells has been shown in wheat protoplasts (Hayashi and Shimamoto, 1988), and feeder cells have been used for maize protoplasts to increase plating efficiency (Rhodes et al., 1988). Plant regeneration from protoplast-derived rice calli is often through somatic embryogenesis (Abdullah et al., 1986; Kyozuka et al., 1987), and active formation of shoots as well as roots has been noted. In particular, vigorously growing roots make survival of regenerated plantlets easy. Somaclonal variation in protoplast-derived rice plants has been extensively studied during the last several years (Ogura et al., 1987, 1989; Nishibayashi et al., 1989). The extent of somaclonal variation is significant, and often morphological changes and low fertility are observed in selfed progeny of protoplast-derived plants. However, the degree of variation varies greatly between regenerated plants and a fraction of them are relatively normal. Also, the degree of variation is dependent on the cultivar, thus the choice of cultivar is important for studies in gene transfer.

Extensive studies on rice protoplast culture by several independent groups (reviewed, in Kyozuka et al., 1989) indicated that establishment of good quality embryogenic suspension cultures is the most important factor for successful plant regeneration from protoplasts. By using the same approach, fertile plants were recently regenerated from maize protoplasts (Shillito et al., 1989; Prioli and Söndahl, 1989), and plantlets have been obtained in wheat (Hayashi and Shimamoto, 1988). These studies should, in the future, lead to the successful production of transgenic maize and wheat plants.

III. Gene Transfer in Rice

Production of transgenic plants has been reported in several cereal species (Table 1). From this it can be concluded that the establishment of reliable protoplast culture is a prerequisite for the production of transgenic cereals because at present, the only method which gives reproducible results in different laboratories is direct DNA transfer to protoplasts. Although high-velocity microprojectiles for delivering DNA into intact cells have been giving encouraging results with cereal cells (Klein et al., 1988; Wang et al., 1988; Mendel et al., 1989), it remains to be seen whether this method can be applied to generate transgenic cereals since somatic cells are not generally competent for cell division or morphogenesis.

Table 1. Transgenic cereals described in the literature

Method	Plant	Transgene	Presence of DNA	Transmission to progeny	Reference
Protoplast (EP)	Maize	*nptII*	+	–	Rhodes et al. (1988)
	Rice	*nptII*	+	?	Toriyama et al. (1988)
	Rice	*nptII*	+	?	Zhang et al. (1988)
	Rice	*hph, gus A*	+	+	Shimamoto et al. (1989)
Protoplast (PEG)	Rice	*gus A*	+	?	Zhang and Wu (1988)
	Orchard grass	*hph*	+	?	Horn et al. (1988)
Injection to inflorescence	Rye	*nptII*	+	?	de la Peña et al. (1987)
Pollen–tube pathway	Rice	*nptII*	+	?	Luo and Wu (1988)

EP, Electroporation; PEG, polyethylene glycol; *nptII*, neomycin phosphotransferase; *hph*, hygromycin phosphotransferase; *gus A*, β-glucuronidase

Because direct DNA transfer into rice protoplasts has been the most extensively studied example in cereal transformation, parameters critically influencing the efficiency of production of transgenic plants are examined below.

A. Method of Introduction

Under optimized conditions, both polyethyleneglycol (Zhang and Wu, 1988) and electroporation (Toriyama et al., 1988; Zhang et al., 1988; Shimamoto et al., 1989) are able to introduce foreign DNA into rice chromosomes. Because transformation frequency is influenced by a number of other factors such as kinds of selectable markers, promoters used for the selectable marker, cell lines. etc., it is not possible to give a definite answer as to which method is better for direct DNA uptake in rice cells. We have developed a protocol for rice transformation by using electroporation (Fig. 1). According to our studies with eight japonica varieties, "competence" or efficiency of DNA uptake by rice protoplasts is variable between varieties and sometimes between different suspension cultures derived from a single cultivar. Non-morphogenic protoplasts isolated from established cell lines are generally more competent than embryogenic protoplasts. To develop a highly efficient method of transformation, it seems necessary to understand the physiology of protoplasts when they are subjected to an electric pulse or PEG treatment.

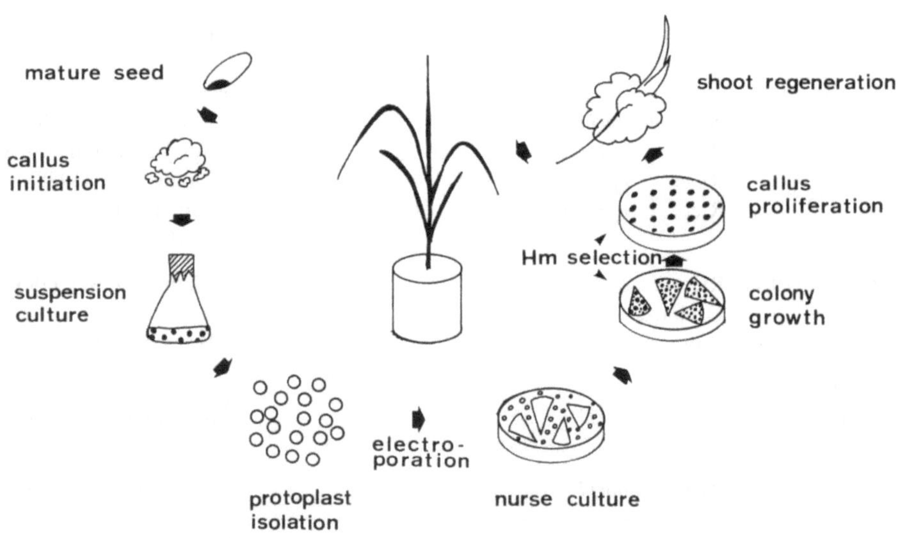

Fig. 1. Scheme of direct gene transfer in rice by electroporation of protoplasts

B. Selectable Marker

Both kanamycin, Km^r (*npt II*), and hygromycin, Hm^r (*hph*), resistance genes have been successfully used as selectable markers to obtain transgenic rice plants (Table 1). When *nptII* is used, kanamycin (Zhang et al., 1988) as well as G418 (Toriyama et al., 1988) have been used to select transformed rice cells. However, effective selection of transformed resistant cells is not always straightforward in rice. The reasons are that, (1) endogenous antibiotic resistance makes the selection difficult, (2) expression of resistance genes is not sufficient to confer a clear resistant phenotype, (3) low survival of resistant cells due to the presence of surrounding sensitive cells. Even if transformed calli can be selected, their regeneration ability or development of plants is sometimes influenced by the process of selection with antibiotics. It has been shown that transformed calli selected in the presence of Km often produce albino plants. It might be concluded that Hm^r is the most suitable selectable marker generally available at this moment for rice cells. It is anticipated, however, that other selectable markers (Dekeyser et al., 1989) will be developed in the future that will make it possible to use combinations of different markers in rice cells.

IV. Integration of Foreign DNA into Rice Chromosomes and Its Transmission to Progeny

Direct DNA transfer often produces rearranged integrated copies in the plant genome (Paszkowski et al., 1984; Rhodes et al., 1988). To use this method for routine production of transgenic plants, patterns of integration of foreign DNA into chromosomes should be examined. For this purpose we have analyzed six Hm^r transformants by Southern blot analysis using various fragments of the Hm^r vector as hybridization probe (Fujimoto et al., unpubl. results). The results show that there are three different types of integration patterns when circular DNA was introduced into rice protoplasts by electroporation: (1) the plasmid was cut at one site and the undegraded molecule was integrated (Fig. 2), (2) up to three fragments derived from different regions of the plasmid are integrated at different sites, and (3) a tandemly-repeated complete plasmid was integrated. With regard to the number of the Hm^r genes present in the rice genome, one to two copies were detected, except in the case of the tandem repeat, which consists of 4–5 copies. This relatively simple pattern of integration of foreign DNA might be due to the low concentration of plasmid DNA (10 μg/ml) used. Higher concentration of plasmid DNA and use of linearized DNA might give different patterns of integration.

Transmission and expression of the Hm^r and *gus A* gene has been demonstrated in the progeny of primary transformants (Shimamoto et al.,

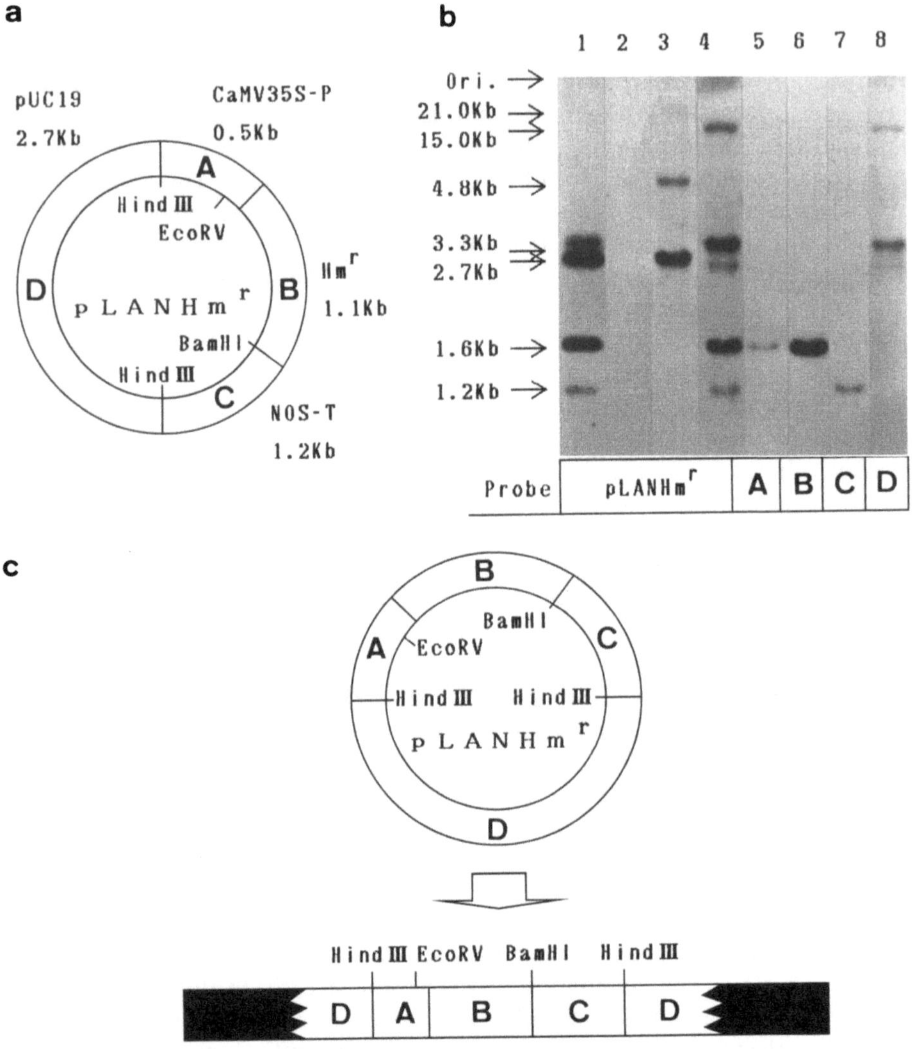

Fig. 2. An integration pattern of the Hm^r plasmid in rice chromosomes. a, Diagram of the Hm^r vector, pLAN Hm^r, and probes A–D used for Southern blot analysis. b, Southern blot analysis of a rice transformant. 1, plasmid pLAN Hm^r digested with *Hind*III and *Bam*HI; 2, DNA of non-transformed rice; 3, DNA of the transformant digested with *Hind*III; 4–8, DNA of the transformant digested with *Hind*III and *Bam*HI. The probes used are shown at the bottom of the blot. c, The structure of the integrated Hm^r plasmid in the rice chromosome deduced from the results of the Southern blot analysis

1989). The GUS marker is an especially convenient marker to examine transmission of transgene to progeny, because its expression can be detected in immature seeds (see Sect. VI). Also, segregation of the integrated copies of introduced DNA has been shown by Southern blot analysis of the progeny

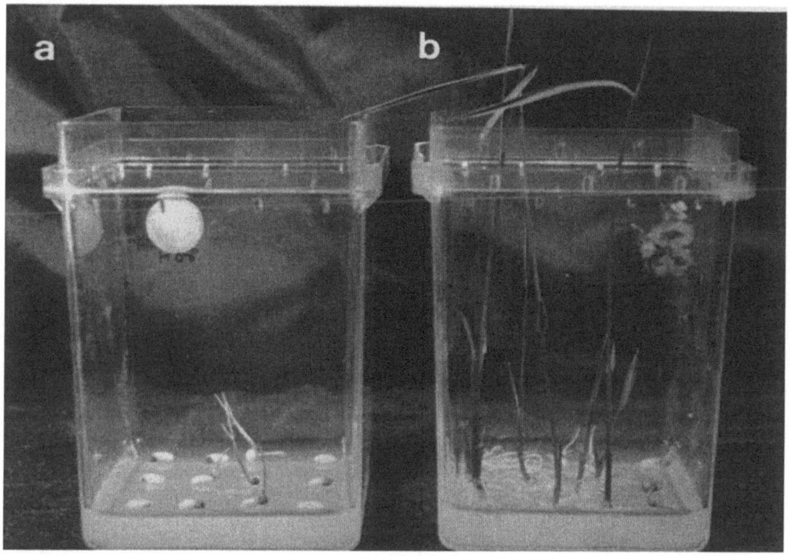

Fig. 3. Expression of the Hm^r phenotype in the selfed progeny of a transgenic rice plant. The germination medium contained 30 μg/ml hygromycin B. a, Untransformed control. b, Selfed progeny of the Hm^r transformant showing segregation of the Hm^r phenotype

obtained from the transgenic plants. That Hm^r progeny seeds exhibit clear Hm^r phenotypes when germinated in the presence of hygromycin B (Fig. 3) indicates that Hm^r is a suitable marker for screening transformants at the seed level.

V. Co-transformation as a Simple Method to Introduce Non-selectable Genes

Co-transformation of unlinked non-selected genes (Schocher et al., 1986) with the Hm^r selectable marker gene was examined by using the *gus A* vector as a scorable marker (Fig. 4). Results showed that the efficiency of co-transformation (frequency of GUS-positive transformants in Hm^r transformants) is 30–50% in a number of experiments with non-embryogenic as well as with embryogenic rice protoplasts (Table 2) (Kyozuka et al., unpubl. results). Because, in direct gene transfer, transforming DNA is likely to be subjected to nuclease action, the size of the integrated DNA may be a limitation. Therefore, construction of a composite plasmid carrying both the selectable marker and genes of interest may not be advantageous especially when the non-selectable gene is of large size. Another merit of co-transformation is that there is no need to combine constructions into a

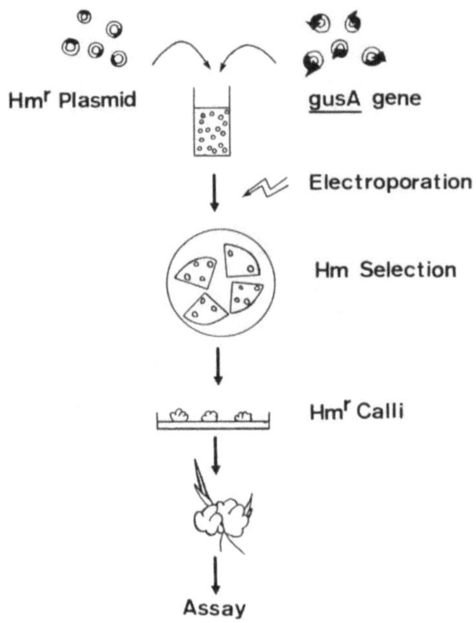

Fig. 4. Procedure for co-transformation of the GUS gene with the Hm^r marker

Table 2. Frequency of co-transformation[a]

Expt.	No. Hm^r calli	No. GUS$^+$ calli	Co-transformation (%)
1	58	30	51.7
2	38	15	39.5

[a] 50 µg/ml each of Hm^r and GUS plasmid was used

single vector, which makes experiments simple when a series of constructs are to be examined.

VI. Expression of Foreign Genes in Transgenic Rice Plants

Although a number of inducible and tissue-specific genes have been examined for their expression in transgenic plants (Benfey and Chua, 1989), only a small number of monocot genes have so far been analyzed because some of them are not expressed at all in transgenic dicot plants or when expressed the expression is not quantitatively or qualitatively correct. Therefore, precise analysis of regulated expression of monocot genes in transgenic plants has been largely precluded because of the lack of a reliable

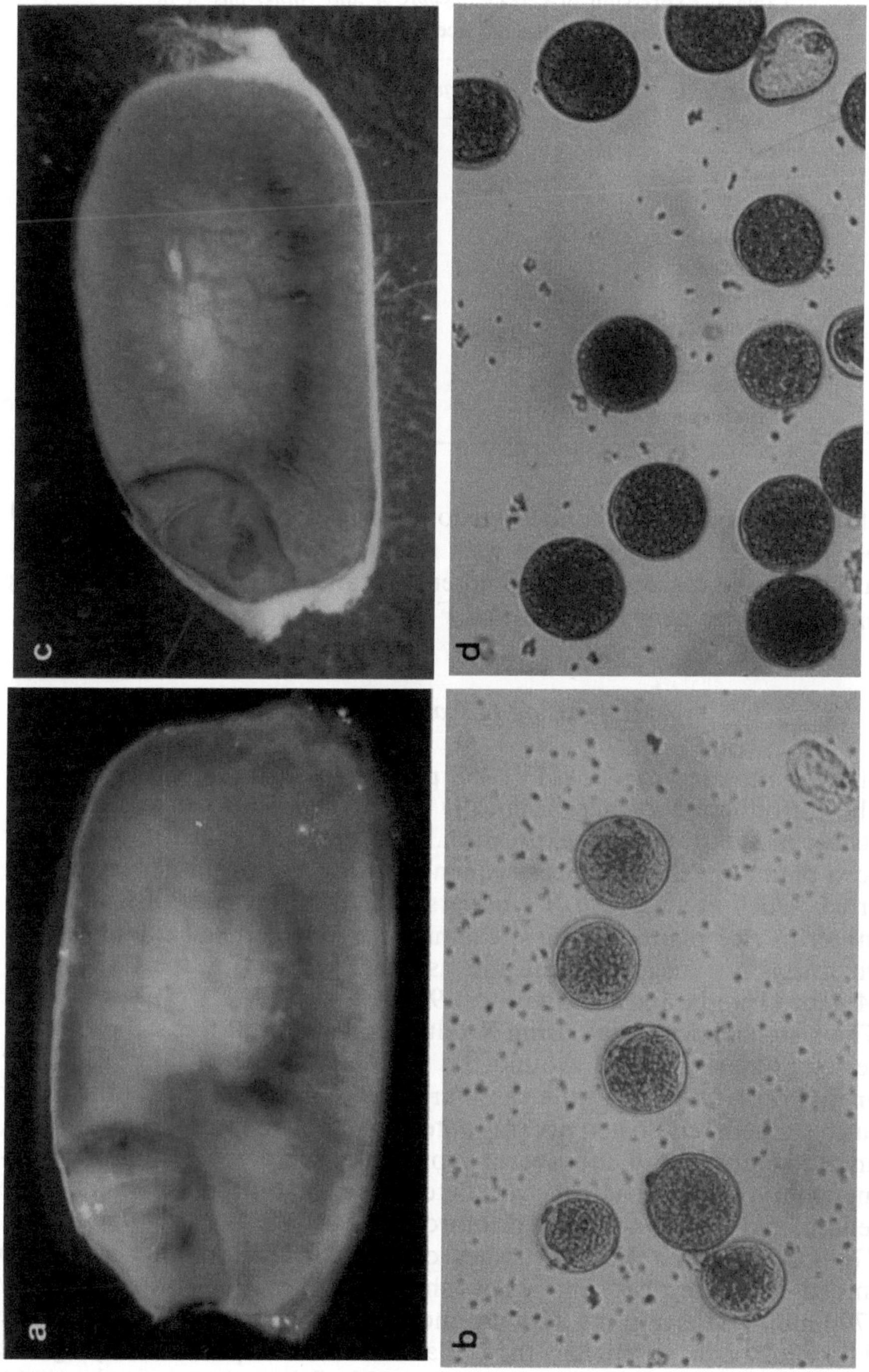

Fig. 5. Expression of CaMV35S-*gus A* and maize *Adh-1-gus A* in seeds and pollen of transgenic rice plants. a and b, CaMV35S-*gus A*; c and d, *Adh-1-gus A*

Table 3. Expression of CaMV35S-*gus A* and *Adh-1-gus A* in transgenic rice plants

Tissue	CaMV35S-*gus A*	*Adh-1-gus A*
Root	+ +	+ + +
Leaf	+ + +	+
Flower		
Anther	−	+ +
Filament	−	+ + +
Pollen	−	+ +
Ovary	−	−
Seed		
Embryo	+ +	+ + +
Endosperm	+	+ +

transformation system in monocotyledonous species. As a first step toward establishing such a model system with rice, we have examined expression of CaMV35S and maize *Adh-1* promoters in transgenic rice plants by using the *gus A* reporter gene (Table 3 and Fig. 5).

A. Expression of CaMV35S-gus A

Cauliflower mosaic virus (CaMV) 35S promoter has been effectively used for the introduction of agronomically useful genes into dicotyledonous plants because of its efficient expression. Also, detailed studies in transgenic tobacco plants on the enhancer sequence within the promoter have been reported (Odell et al., 1985). To understand the properties of CaMV35S promoter in rice plants, we have examined by histochemical and fluorometric assays a transgenic plant and its progeny carrying the CaMV35S-*gus A* gene (Terada and Shimamoto, 1990).

The histochemical assay using X-Gluc indicated that the 35S promoter expresses primarily in and around the vascular bundle of the leaf, flower organ and root. Other than the vascular bundle, the basal part of the ovary was also stained, but no activity was detected in pollen grains. Interestingly, the mature embryo and endosperm expressed GUS activity, and in germinating embryos, a high activity was detected in the outermost layers of the scutellum, and in leaf and root primordia.

GUS activities in various tissues of the transgenic plants were also examined. Mature leaf and root showed a relatively high activity (20–70 nmol 4-MU/min/mg protein) indicating that the leaf GUS activity level is similar to or higher than that of tobacco plants carrying the CaMV35S-*gus A* gene (Jefferson et al., 1987; Schernthaner et al., 1988).

These results suggest that the CaMV35S promoter is as effective in rice plants as in tobacco and that the CaMV35S promoter should be useful for introduction of a variety of foreign genes into plants.

B. Expression of Maize Adh-1 Promoter

Maize alcohol dehydrogenase-1 (*Adh-1*) is one of the best studied genes in higher plants (Freeling and Bennet, 1985). In maize, the product of *Adh-1* is present in seed, seedling and pollen. The expression of *Adh-1* is induced in roots in response to anaerobic conditions and its induction is regulated at the transcriptional level (Vayda and Freeling, 1986). In order to examine whether the promoter activity of a maize gene is correctly regulated in another graminacious species, rice, the promoter was fused with *gus A* gene and analyzed for its expression in various tissues of transgenic rice plants (Kyozuka et al., 1990; Kyozuka et al., unpubl. results). In the plant, GUS was predominantly expressed in roots and flower tissues, and both the embryo and endosperm of the seed contained considerable GUS activity. Clear staining of anther, anther filament and pollen from the plant carrying the *Adh-1-gus A* construct was distinguished from that of the CaMV35S-*gus A*, in which very little activity was observed in these tissues (Table 3 and Fig. 5). From the histochemical study of the rice plants expressing maize *Adh-1* promoter, it was concluded that its expression is correctly regulated in rice suggesting that transgenic rice plants would be suitable for the study of the regulation of monocotyledonous genes.

VII. Discussion

Establishment of reproducible plant regeneration from protoplasts has made rice a model cereal species with which to apply various types of genetic manipulation. In particular, establishment of reliable gene transfer methods in this species opens possibilities of genetic engineering for future breeding as well as for studying regulation of monocot genes using transgenic plants.

Because protoplast-derived plants are used for various studies, examination of somaclonal variation (mutation induced during culture) present in those plants is becoming increasingly important. We have examined R_1 population of protoplast-derived plants from more than 10 cultivars during the last several years. The degree of variation found is significant in some cultivars, but not in others when gross phenotype was examined. Often plants with low fertility were detected. Whether the mutations found in protoplast-derived plants will seriously influence their use for crop improvement remains to be studied, however, it is safe to point out that when a reasonable (10 or more) number of transgenic plants are examined, those in which minimum amounts of mutation had accumulated could be found.

One of the shortcomings of direct gene transfer is that transgenes are often cut at various regions, and that no reasonable method of preventing this exists at present. Linearization at sites flanking the transgene might be a solution although there is an indication that it leads to concatenation of plasmid DNA, making interpretation of expression of foreign genes complicated (Riggs and Bates, 1986). In particular, studies aiming for the dissection of promoter regions, as has been done with a number of genes using Ti-mediated transformation, are difficult to perform with the direct DNA transfer method unless the integrated DNA is precisely examined by extensive Southern blot analysis or polymerase chain reaction. To improve the available gene transfer technique in cereals, further studies to identify types of transforming DNA which are less sensitive to nuclease action or to develop novel vectors that could transfer defined segments of DNA will be required.

Establishment of a model monocot system to examine regulation of monocot genes in transgenic plants is urgently required for the study of their regulation. To assess suitability of rice as such a system, the promoters of CaMV35S and maize *Adh-1* were fused with the *gus A* reporter gene and introduced into rice plants. In transgenic rice plants, the CaMV35S promoter is expressed primarily in vascular bundles of roots, leaf and flower organs. No expression was detected in most of the flower tissues and pollen. This is basically similar to the results obtained with transgenic tobacco transformed with CaMV35S-*gus A* gene. The level of expression in rice was found to be similar to that in tobacco. The maize *Adh-1* promoter does not express at all in transgenic tobacco plants unless the enhancer sequence of octopine synthase or CaMV35S was inserted upstream of the promoter region (Ellis et al., 1987). On the contrary, in transgenic rice plants the *Adh-1* promoter was primarily expressed in roots, flower tissues and seed but very little in the leaf. Most interestingly, the promoter is strongly expressed in the root cap which is in contrast to the CaMV35S promoter that is strongly expressed in those cells surrounding the division center of the root. Also, the *Adh-1* promoter is efficiently expressed in pollen grains and most of the flower tissues. These results suggest that expression of the maize *Adh-1* promoter in transgenic rice plants is regulated as in maize and that rice can be used to study the regulation of maize genes.

Production of transgenic rice plants is now routine, and prospects of developments in transformation of other major cereals are also good. In maize, reports of production of fertile plants from protoplasts should be followed in the near future by generation of fertile transgenic plants, because all the conditions necessary for transformation of maize protoplasts and plant regeneration have been established (Rhodes et al., 1988). In wheat and barley, however, a significant progress in plant regeneration from protoplasts has not yet been demonstrated, although there are some reports on regeneration of abnormal plants from protoplasts. Use of high-velocity microprojectiles is an attractive alternative to the use of protoplasts in cereal transformation. However, it is well known that differentiated somatic cells of cereal species are not generally competent for cell division or morphogenesis. Moreover, it has been shown to be extremely difficult to establish embryogenic suspensions in wheat and barley compared to rice. Thus, at the moment, the only tissue which might be effectively used to apply this technique in these species are embryos. Whether high-velocity microprojectiles can be used with cereal embryos, and whether foreign DNA can enter the germ line cells of the embryo remains to be studied.

Acknowledgement

I thank J. Kyozuka, H. Fujimoto and T. Izawa for providing their unpublished results. I also thank R. Terada, Y. Hayashi and M. Nakajima and other members of Plantech Research Institute for helpful discussions.

VIII. References

Abdullah R, Cocking EC, Thompson JA (1986) Efficient plant regeneration from rice protoplasts through somatic embryogenesis. Bio/Technology 4: 1087–1090

Benfey PN, Chua N-H (1989) Regulated genes in transgenic plants. Science 244: 174–181

Bennett MD, Smith JB, Heslop-Harrison JS (1982) Nuclear DNA amounts in angiosperms. Proc R Soc Lond [Biol] 216: 179–199

de la Pẽna A, Lörz H, Schell J (1987) Transgenic rye plants obtained by injecting DNA into young floral tillers. Nature 325: 274–276

Ellis JG Llewellyn DJ, Dennis ES, Peacock WJ (1987) Maize *Adh-1* promoter sequences control anaerobic regulation: addition of upstream promoter elements from constitutive genes is necessary for expression in tobacco. EMBO J 6: 11–16

Freeling M, Bennett DC (1985) Maize *Adh-1.* Annu Rev Genet 19: 297–323

Gasser CS, Fraley RT (1989) Genetically engineering plants for crop improvement. Science 244: 1293–1299

Hayashi Y, Shimamoto K (1988) Wheat protoplast culture: embryogenic colony formation from protoplasts. Plant Cell Rep 7: 414–417

Horn ME, Shillito RD, Conger BV, Harms CT (1988) Transgenic plants of orchardgrass (*Dactylis glomerata* L.) from protoplasts. Plant Cell Rep 7: 469–472

Keith B, Chua N-H (1986) Monocot and dicot pre-mRNA are processed with different efficiencies in transgenic tobacco. EMBO J 5: 2419–2425

Klein TM, Fromm M, Weissinger A, Tomes D, Schaaf S, Sletten M, Sanford JC (1988) Transfer of foreign genes into intact maize cells with high-velocity microprojectiles. Proc Natl Acad Sci USA 85: 4305–4309

Kyozuka, J, Hayashi Y, Shimamoto K (1987) High frequency plant regeneration from rice protoplasts by novel nurse culture methods. Mol Gen Genet 206: 408–413

Kyozuka J, Otto E, Shimamoto K (1988) Plant regeneration from protoplasts of indica rice: genotypic differences in culture response. Theor Appl Genet 76: 887–890

Kyozuka J, Shimamoto K, Ogura H (1989) Regeneration of plants from rice protoplasts. In: Bajaj YSP (ed) Biotechnology in agriculture and forestry, vol 8, plant protoplasts and genetic engineering I. Springer, Berlin Heidelberg New York Tokyo, pp 109–123

Kyozuka J, Izawa T, Nakajima M, Shimamoto K (1990) Effect of the promoter and the intron of maize *Adh-1* on foreign gene expression in rice. Maydica 35: 1–5

Luo ZX, Wu R (1988) A simple method for the transformation of rice via the pollen-tube pathway. Plant Mol Biol Rep 6: 165–174

Mendel RR, Muller B, Schulze J, Kolesnikov V, Zelenin A (1989) Delivery of foreign genes into intact barley cells by high-velocity microprojectiles. Theor Appl Genet 78: 31–34

Nishibayashi S, Hayashi Y, Kyozuka J, Shimamoto K (1989) Chromosome variation in protoplast-derived calli and in plants regenerated from the calli of cultivated rice (*Oryza sativa* L.). Jpn J Genet 64: 355–361

Odell JT, Nagy F, Chua N-H (1985) Identification of DNA sequences required for activity of the cauliflower mosaic virus 35S promoter. Nature 313: 810–812

Ogura H, Kyozuka J, Hayashi Y, Shimamoto K (1987) Field performance and cytology of protoplast-derived rice (*Oryza sativa*): high yield and low degree of variation of four japonica cultivars. Theor Appl Genet 74: 670–676

Ogura H, Kyozuka J, Hayashi Y, Shimamoto K (1989) Yielding ability and phenotypic trait in the selfed progeny of protoplast-derived rice plants. Jpn J Breed 39: 47–56

Paszkowski J, Shillito RD, Saul M, Mandak V, Hohn T, Hohn B, Potrykus I (1984) Direct gene transfer to plants. EMBO J 3: 2717–2722

Prioli LM Söndahl MR (1989) Plant regeneration and recovery of fertile plants from protoplasts of maize (*Zea mays* L.). Bio/Technology 7: 589–594

Rhodes CA, Pierce DA, Mettler IJ, Mascarenhas D, Detmer JJ (1988) Genetically transformed maize plants from protoplasts. Science 240: 204–207

Riggs CD, Bates GW (1986) Stable transformation of tobacco by electroporation: evidence for plasmid concatenation. Proc Natl Acad Sci USA 83: 5602–5606

Robert LS, Thompson RD, Flavell RB (1989) Tissue-specific expression of a wheat high molecular weight glutenin gene in transgenic tobacco. Plant Cell 1: 569–578

Schell JS (1987) Transgenic plants as tools to study the molecular organization of plant genes. Science 237: 1176–1183

Schernthaner JP, Matzke MA, Matzke AJM (1988) Endosperm-specific activity of a zein gene promoter in transgenic tobacco plants. EMBO J 7: 1249–1255

Schocher RJ, Shillito RD, Saul MW, Paszkowski J, Potrykus I (1986) Co-transformation of unlinked foreign genes into plants by direct gene transfer. Bio/Technology 4: 1093–1096

Shillito RD, Carswell GK, Johnson CM, DiMaio JJ, Harms CT (1989) Regeneration of fertile plants from protoplasts of elite inbred maize. Bio/Technology 7: 581–588

Shimamoto K, Terada R, Izawa T, Fujimoto H (1989) Fertile transgenic rice plants regenerated from transformed protoplasts. Nature 338: 274–276

Terada R, Shimamoto K (1990) Expression of CaMV35S-GUS gene in transgenic rice plants. Mol Gen Genet 220: 389–392

Toriyama K, Arimoto Y, Uchimiya H, Hinata K (1988) Transgenic rice plants after direct gene transfer into protoplasts. Bio/Technology 6: 1072–1074

Vayda ME, Freeling M (1986) Insertion of the *Mu1* transposable element into the first intron of maize *Adh1* interferes with transcript elongation but does not disrupt chromatin structure. Plant Mol Biol 6: 441–454

Wang YC, Klein TM, Fromm M, Cao J, Sanford JC, Wu R (1988) Transient expression of foreign genes in rice, wheat and soybean cells following particle bombardment. Plant Mol Biol 11: 433–439

Zhang W, Wu R (1988) Efficient regeneration of transgenic plants from rice protoplasts and correctly regulated expression of the foreign gene in the plants. Theor Appl Genet 76: 835–840

Zhang HM, Yang H, Rech EL, Golds TJ, Davis AS, Mulligan BJ, Cocking EC, Davey MR (1988) Transgenic rice plants produced by electroporation-mediated plasmid uptake into protoplasts. Plant Cell Rep 7: 379–384

Chapter 2

Genetic Transformation of Potato to Enhance Nutritional Value and Confer Disease Resistance

Luis Destéfano-Beltrán[1], Pablito Nagpala[1], Kim Jaeho[1], John H. Dodds[2], and Jesse M. Jaynes[1]

[1]Department of Biochemistry, 322 Choppin Hall, Louisiana State University, Baton Rouge, LA 70803, U.S.A.
[2]International Potato Center, P.O. Box 5969, Lima, Peru

With 9 Figures

Contents

I. Introduction

The potato is one of the most important calorie and protein sources in many developed and developing countries with total production yielding about 95 million tons of tubers worth about $ 24 billion dollars (1984 figures). The nutritional quality of the potato tuber protein, although relatively high, is, like for most plant proteins, deficient in certain essential amino acids, e.g. lysine and methionine. The expression of synthetic genes encoding proteins rich in essential amino acids, along with normal protein production within the tuber, may increase the overall nutritional quality of the potato as well as its aggregated value in the market especially for the food industry.

In the last 15 years there has been a tremendous effort expended to introduce new varieties appropriate for the tropical areas (Sawyer, 1984). One primary component in restricting potato production in those regions is the presence of bacterial and fungal diseases. Losses in the worldwide production of potato associated with bacterial diseases can be as high as twenty-five percent especially in very bad years. The incorporation of genes into plants encoding potent antimicrobial proteins, derived from insects, may significantly augment the level of their resistance to bacterial and fungal disease. Using recently developed protocols, genes of choice can now be routinely introduced into the genome of potato plants.

Fortunately, the potato has many advantages over other major crop plants. It is easily manipulated in tissue culture, both in regeneration and propagation methods, and is susceptible to infection by *Agrobacterium tumefaciens* and *A. rhizogenes*. In this chapter, we will describe our work on the use of genetic engineering in potato for nutritional quality improvement and enhancement of bacterial and fungal disease resistance.

II. Nutritional Improvement

The biosynthesis of amino acids from simpler precursors is a process vital to all forms of life as these amino acids are the building blocks of proteins. Organisms differ markedly with respect to their ability to synthesize amino acids. In fact, virtually all members of the animal kingdom are incapable of manufacturing some amino acids. There are twenty common amino acids which are utilized in the fabrication of proteins and essential amino acids are those protein building blocks which cannot be synthesized by the animal. It is generally agreed that humans require eight of the twenty amino acids in their diet. Protein malnutrition can usually be ascribed to an intake which is deficient in one or more of these essential amino acids and their daily consumption is a requisite for a nutritionally adequate diet.

When diets are high in carbohydrates and low in protein, over a protracted period, essential amino acid deficiencies result. The name given to this undernourished condition is "Kwashiorkor" which is an African word meaning "deposed child" (deposed from the mother's breast by a newborn sibling). This debilitating and malnourished state, characterized by a bloated stomach and reddish-orange discolored hair, is more often found in children than adults because of their great need for essential amino acids during growth and development. In order for normal physical and mental maturation to occur, the above mentioned daily source of essential amino acids is a requirement. Essential amino acid content, or protein quality, is as important a feature of the diet as total protein quantity or total calorie intake. Our protein quality work began with a rat feeding experiment we conducted utilizing polylysine as a source of the essential amino acid lysine. What we found was that this polymer could be utilized as an efficient source of free lysine in the diet of the rat (Newman et al., 1980). The data we reported indicated the great importance an adequate diet of essential amino acids plays in normal animal development and overall physical well-being.

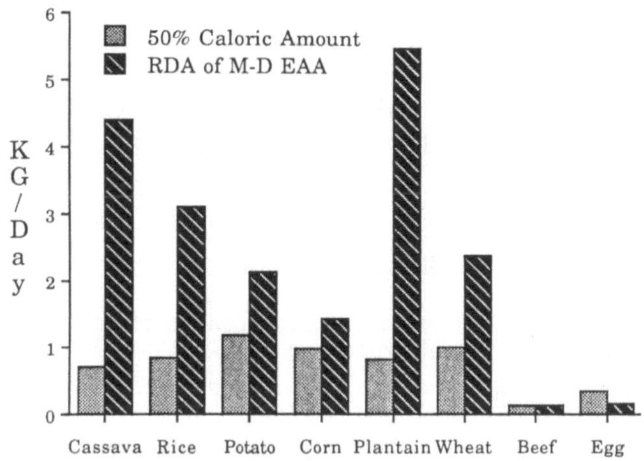

Fig. 1. Necessary daily consumption of some of the most important animal and plant protein sources in kg per day to meet 50% of the daily caloric intake and the required daily amount for the five most-deficient essential amino acids

Some foods, such as milk, eggs, and meat, have very high nutritional values because they contain a disproportionately high level of essential amino acids. On the other hand, most foodstuffs obtained from plants possess a poor nutritional value because of their relatively low content of some or, in a few cases, all of the essential amino acids (Fig. 1). In general, the essential amino acids which are found to be most limited in plants are isoleucine, lysine, methionine, threonine, and tryptophan.

In the diet of people inhabiting a typical developed country, few amino acid deficiency problems arise. Primarily, because their diet is composed of a mixture of a wide range of animal and plant proteins. That situation is, however, not true in many developing countries. In a number of cases, the total food intake, perhaps 80–90%, is highly dependent on a single crop. Rice, for example, is the major staple in Asia while potatoes are the staple in the Andean region of South America. Once there is heavy dependence on plant protein from a single source, its essential amino acid composition becomes of critical importance. In those situations, where the food source is a single plant, it would be highly beneficial to "engineer" the plant to produce proteins with a balanced essential amino acid content.

Our initial strategy to challenge this problem was to design a synthetic protein that would encode a random "polymer" of essential amino acids similar to the polylysine. This first designed protein is about 13 kDa, with a content of up to 80% of essential amino acids. A synthetic fragment, 292 bp long, encoding this protein was constructed, cloned, and expressed in bacteria (Jaynes et al., 1985). This novel gene fragment, HEAAE-DNA

(High Essential Amino Acid Encoding DNA), as a result of the gene design, codes actually for two proteins with a high content of essential amino acids found to be most deficient in plant-derived protein (Hansen, 1979). The sequence of HEAAE-DNA and resultant protein sequences are shown (Fig. 2). Proteins A and B represent sequences derived from both possible

Reading Direction of Bottom Strand (A) →

```
AATTCGGGGATCGTAAGAAATGGATGGATCGTCATCCATTTCTTCATCCATTTCTTACGA
TTAAGCCCCTAGCATTCTTTACCTACCTAGCAGTAGGTAAAGAAGTAGGTAAAGAATGCT

TCCATCCATTTCTTAAGAAATGGATGAAGAAATGGATGACGATCCATCCATTTCTTCATC
AGGTAGGTAAAGAATTCTTTACCTACTTCTTTACCTACTGCTAGGTAGGTAAAGAAGTAG

CATTTCTTCATCCATTTCTTACGATCAAGAAATGGATGAAGAAATGGATGAAGAAATGGA
GTAAAGAAGTAGGTAAAGAATGCTAGTTCTTTACCTACTTCTTTACCTACTTCTTTACCT

TGAAGAAATGGATGCATCCATTTCTTAAGAAATGGATGAAGAAATGGATGAAGAAATGGA
ACTTCTTTACCTACGTAGGTAAAGAATTCTTTACCTACTTCTTTACCTACTTCTTTACCT

TGACGATCGATCGTAAGAAATGGATGACGATCCATCCATTTCTTACGATCCCCGAATT
ACTGCTAGCTAGCATTCTTTACCTACTGCTAGGTAGGTAAAGAATGCTAGGGGCTTAA
```

← Reading Direction of Bottom Strand (B)

Sequence of Protein A

GDR**KKWM**DRHP**FL**HP**FLTI**HP**FLKKWMKKWMTI**HP**FL**HP

FLHP**FLTIKKWMKKWMKKWMKKWM**HP**FLKKWMKKWMKKW**

MTIDR**KKWMTI**HP**FLTI**P

Sequence of Protein B

GDR**KKWM**DRHP**FLTI**DRHP**FL**HP**FL**HP**FLKKWM**HP**FL**HP

FLHP**FL**HP**FL**DR**KKWMKKWMKKWM**DRHP**FL**HP**FLKKWM**D

R**KKWMKKWMTI**HP**FLTI**P

Fig. 2. The nucleotide sequence of HEAAE I DNA and the derived protein sequences. This DNA yields two different proteins designated A and B. One or the other will be produced depending upon the orientation of the gene when it is fused to the CAT protein under the control of a suitable plant promoter. Protein A, the best so far analyzed, is composed of about 80% essential amino acids (those which are bold)

reading directions of the gene. The compositions of proteins A and B, considering the most deficient essential amino acids of plant-derived protein, are three-fold higher than those found in milk or egg protein (Hansen, 1979). The 292 bp fragment was subcloned in the only EcoRI site of the plasmid pGA414, which has the chloramphenicol acetyl transferase (CAT) gene under the transcriptional control of the nopaline synthetase (NOS) promoter and the 3′ end of NOS for polyadenylation. This chimeric CAT-HEAAE gene was subsequently transferred to pFW105 which possesses a fragment of T-DNA for recombination purposes and wide host range properties and can replicate in *Agrobacterium rhizogenes*. The recombinant constructs, pFW105.10.2 and pFW105.10.3, were introduced into the agropine-producing strain R1000 of *A. rhizogenes* by direct transfer and double recombinants were selected on appropriate antibiotics.

In vitro-derived plants were wounded and infected with various *A. rhizogenes* clones. Hairy roots formed within two weeks and were excised two weeks later to be transferred to regeneration media. Tubers were induced in vitro on regenerated plantlets following standard techniques (Espinoza et al., 1984). In general, mini-tubers (3–5 mm in diameter) were obtained after 8 to 10 weeks. Southern and Northern analysis showed the proper integration of the constructs and expression of the chimeric mRNA respectively. Western blots of total tuber-protein confirmed the presence of a protein of the expected size, 38 kDa. This first attempt did not result in any significant increase in the levels of essential amino acids, as indicated by amino acid analysis of total tuber-protein. We think this was due to a low level of expression, instability of the protein and the fact that the HEAAE-protein was "diluted" in the CAT-HEAAE fusion protein (Yang et al., 1989). The same gene was introduced into tobacco by *A. tumefaciens* mediated-transformation with similar results when total leaf-protein was analyzed for any increase in essential amino acid content and total levels of protein (Jaynes et al., unpubl. results).

While we obtained detectable levels of this high quality protein in the potato plants, we were not satisfied with our results and decided to design a new protein in a very systematic and logical way. We first looked at the amounts of essential amino acids necessary for normal metabolism of the human being, from infants to the aged. From these data, we deduced a set of numbers which we call the "needs ratio". We also determined the "deficiency values" or the ratios of deficient essential amino acids for the 12 primary crops people consume throughout the world. From these data, we then found the ratio of essential amino acids needed to totally complement each particular plant foodstuff. We merely averaged these values and came up with a set of numbers we call the "Average Ratio for All Crops Idealized to the HEAAE II Monomer". This set of numbers represents the ratio of essential amino acids necessary to complement the deficiencies found in all 12 crops for all human age groups (Fig. 3).

Ile	1.77	→	2.0
Leu	1.87	→	2.0
Lys	3.20	→	4.0
Met	2.90	→	3.0
Phe	1.48	→	1.0
Thr	1.84	→	2.0
Trp	1.53	→	1.0
Val	1.45	→	1.0
			16.0 Total/Monomer

Fig. 3. The numbers represent the idealized ratio of essential amino acids necessary to complement the deficiencies found in the 12 major food crops for all human age groups

From the above set of numbers, we designed the HEAAE II protein. It was constructed in such a way as to be extremely stable with a very high degree of aggregation. It is also most probably very bioavailable, and capable of being expressed to high levels in the plant (bioavailability refers to the amount of amino acids actually absorbed from a particular dietary protein and used by the organism to make its own protein). This was accomplished by utilizing what we have learned in the design of "disease resistance proteins". The HEAAE II protein is extremely high in essential amino acids, much better than egg or milk proteins (Fig. 4).

The protein's ability to aggregate very closely resembles what occurs with natural plant storage proteins (of course, natural plant storage proteins are extremely poor in essential amino acid content). Indeed, in vitro data utilizing a HEAAE II synthetic monomer, has shown that it behaves as a hexamer in gel-filtration studies. Also, circular dichroism analysis of such a monomer, has indicated that the peptide is almost 100% α-helical in salt concentrations up to 6 M (Kim and Jaynes, in prep.). The gene construct for the plant transformation experiments consisted of a tetramer assembled from the stepwise addition of monomer units. We have preliminary data which indicates that we can cause the leaves of transformed tobacco plants expressing HEAAE II to accumulate much more protein—from ~150%—above control plants in the case of some of our transformed lines (Fig. 5). These results, if true, are highly significant and portend that in the near future we may achieve our long-sought goal of producing high quality protein in plants. This would then make plants like eggs, milk, or meat in protein quality.

This method of gene synthesis is flexible enough to produce proteins possessing any particular amino acid composition. Therefore, proteins could be specifically designed to supplement any desired animal feed or human food. It should be pointed out that the insertion of lysine at frequent intervals in these synthetic proteins provides numerous sites for proteolytic attack by trypsin (one of the main protein-degrading enzymes found in the digestive tract). This feature is important as it increases the bioavailability of the supplemental protein.

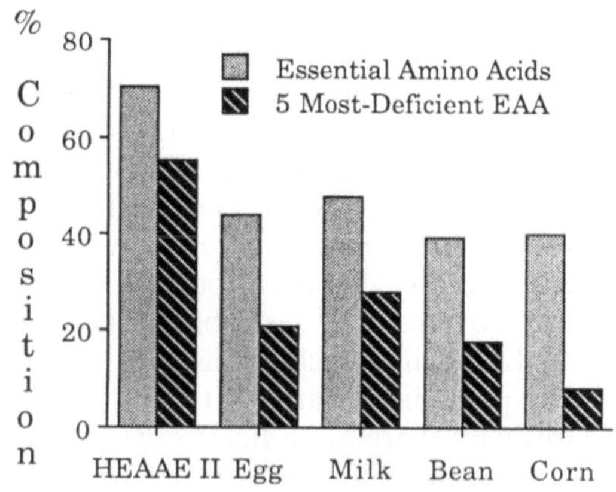

Fig. 4. Comparison of the essential amino acid content of HEAAE II tetramer with those from important human foods. The five most deficient essential amino acids are: isoleucine, lysine, methionine, threonine, and tryptophan

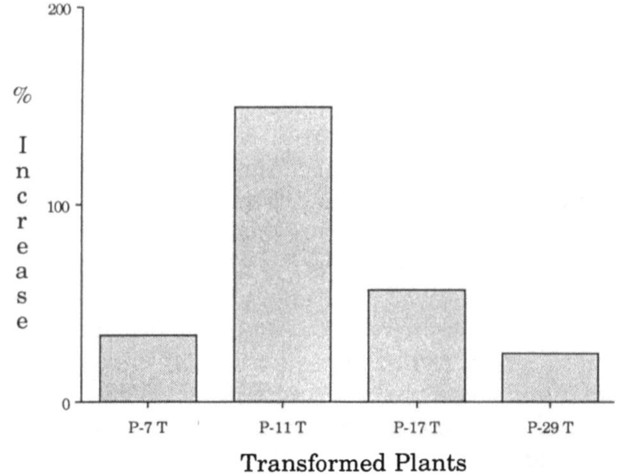

Fig. 5. The data plotted are taken from amino acid analysis of leaves from 4 individual tobacco plants transformed with the gene HEAAE II which codes for a protein high in essential amino acids. Each histogram is the average of the sums of at least three to five separate determinations on an amino acid analyzer

III. Enhancing Disease Resistance in Plants

Since the beginning of history, the well-being of human kind has been unquestionably bound to their ability to cultivate plants efficiently and productively. Plant disease, a disruptive situation, has been at times a catastrophic occurrence altering the lives of millions of people. For instance, late blight, a disease resulting from an infection by a fungal pathogen, caused the starvation of one million people and forced the immigration of another two million to North America due to its decimation of the potato crop (the infamous Irish Potato Famine of 1845–1860). It is estimated that as much as one third of the total crop losses in the world can be directly attributed to plant disease (Agrios, 1978). The loss in the potato, associated with bacterial disease, can be as high as twenty five percent (about US$ 4 billion annually) of the total worldwide production (Sawyer, 1984).

Plant breeding, the movement of desirable traits between plants by the use of the traditional Mendelian techniques, has played a major role in modern agriculture by providing plants with increased disease resistance and higher yields. However, there are inherent problems with this method. For example, it may take as long as ten to fifteen years, to introduce, select, and establish a particular trait into a commercial cultivar and, very often, some traits of agronomic importance are impossible to introduce by these conventional techniques. Thus, in the case of potato, despite the enormous genetic variability available, particularly in the Andean region, efforts in widening the gene pool of cultivated varieties have been relatively unsuccessful. In addition, breeding for disease resistance has been complicated by the continuing emergence of "new" pathogens which eventually overcome the "resistant" variety.

The Peruvian potato (*Solanum tuberosum* L.) is susceptible to a great number of diseases, some of which are of worldwide importance whereas others are of more localized significance. The causal agents include bacteria, fungi, viruses, viroids, mycoplasms, and nematodes (Rich, 1983). There are more numerous diseases in potato grown in the tropics than those cultivated in temperate areas. Indeed, Wellman (1972) reported that "*Solanum andigenum-tuberosum* is afflicted with at least 175 different diseases in the tropics compared to 91 found in the temperate areas". Of those in the United States, ten have been reported to be caused by bacteria and thirty from fungal infections (Anonymous, 1960). Among the most important bacterial diseases are bacterial soft rot, caused by *Erwinia carotovora* and bacterial wilt (Brown Rot), caused by *Pseudomonas solanacearum*. Total losses from these two diseases can be between 30–100% during cultivation and the 2–6 months storage period where temperatures can be between 27–32 °C. The most important fungal diseases are caused by *Rhizoctonia solani*, *Alternaria solani*, *Verticillium dahliae*, and *Phytophthora infestans* (Annual Report, 1988, International Potato Center). Total crop losses in the potato, resulting from bacterial and fungal diseases, can amount to hundreds of millions of dollars (US$) annually in developing countries. During particularly bad years, damages can approach several billion dollars worldwide. The extent of loss varies greatly from country to country and is influenced by climate and conditions of growth and storage. However, in general, losses tend to be greater in developing nations (Fig. 6).

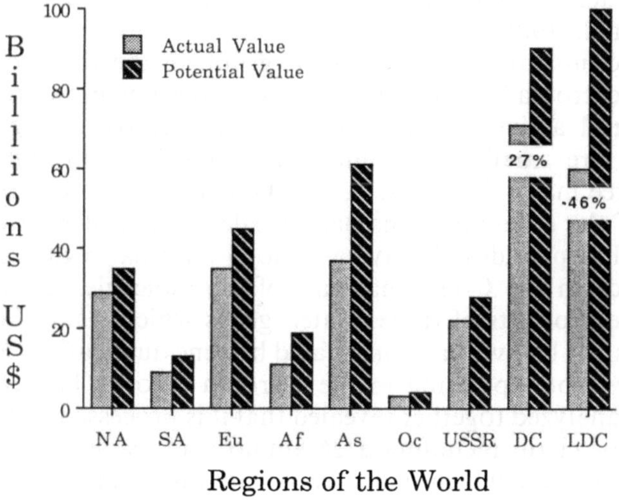

Fig. 6. Histograms showing the economic losses in agricultural output due to disease. Losses are more significant in less developed countries (LDC) than developed countries (DC). NA, North America; SA, South America; Eu, Europe; Af, Africa; As, Asia; Oc, Oceania

Genetic engineering promises to accelerate the breeder's efforts in obtaining plants with an increased resistance to virus, herbicides, and insects (Fraley et al., 1988). Insect-tolerant plants are an example of how genetic engineering has increased the gene pool available to the breeder. Traditionally, breeders relied on the existence of alternative characteristics found among plants of the same or closely related species. By using the insect toxin gene of *Bacillus thuringensis* (B.t.), researchers have opened new doors for crop improvement.

In contrast to insect-tolerant plants engineered with a bacterial insect-toxin, microbial-tolerant plants have been obtained in our lab by using insect-derived lytic peptides.

Humoral immunity can be induced in insects by an injection of either live, non-pathogenic bacteria or heat-killed pathogens. This phenomenon has been studied by many investigators in a number of different insects, especially those of the orders Lepidoptera and Diptera. However, diapausing pupae of the giant silk moth *Hyalophora cecropia*, have proven to be particularly useful (mainly because of their large size) to study the humoral defense mechanisms of insects. When *H. cecropia* pupae are immunized they produce a set of ~15 proteins which are normally not present in the hemolymph of the animals. Among these, there is a set of three (of which two are novel) classes of bactericidal proteins: cecropins, attacins, and a lysozyme (Hultmark et al., 1980). The lysozyme protein is very similar to that found in chicken egg white. The amino acid regions responsible for the catalytic activity and for the binding of substrate are highly conserved (Engstrøm et al., 1985).

The cecropins are the most potent group of the antibacterial factors found in the cecropia hemolymph. These peptides possess a broad spectrum of antibacterial activity against both gram negative and gram positive forms. They are small, 35–37 amino acid residues, strongly basic, and comprise three major forms: A, B, and D. Comparison of the amino acid sequences of the different forms has revealed a high degree of homology ~60–80%. The peptides all have a basic N-terminal region and a hydrophobic stretch in the C-terminal part of the molecule. It seems that the cecropins are products of three related genes which, as in the case of the attacin genes (see below), have originated by gene duplication. Recombinant cDNA clones, corresponding to the cecropin B form, have been isolated which when analyzed together revealed that it is processed from a 62-amino acid precursor chain including a 26-amino acid leader peptide and a C-terminal Gly residue which is decarboxylated to render an amidated Leu (Van Hofsten et al., 1985). Recently, several genomic clones have been characterized and cecropin B was shown to exist in at least 3–5 copies in the cecropia genome (Xanthopoulos et al., 1988).

Additional cecropin-like lytic peptides have been described in *Antheraea pernyi* (Qu et al., 1982), *Bombyx mori* (Shiba et al., 1983), and *Drosophila melanogaster* (Flyg et al., 1987). The sarcotoxins: IA, IB, and IC, were isolated from the hemolymph of *Sarcophaga peregrina* larvae (Okada and Natori, 1983). These peptides, although small and basic, possess a somewhat different structural motif from that exhibited by the cecropins (Jaynes et al., 1990). However, all appear to act by causing generalized membrane disruption.

Attacins are the largest antibacterial molecules found in the hemolymph of immunized *H. cecropia* pupae, with a molecular weight of about 20 kDa. They are comprised of six different isoforms (A through F), which can be fractionated according to their isoelectric point (Hultmark et al., 1983). The results from amino acid sequence analysis of the N-terminus of five of the attacins indicate the presence of three basic and two acidic species. The basic type has similar sequences while the acidic pair are identical. It has been suggested that they are the products of two related genes. Comparison of cDNA clones has revealed ~76% homology in the coding region and is, thus, evidence for the origination of the genes from a common ancestral form. The six attacin isoforms found in the hemolymph are thought to be products of secondary modification of the two precursor molecules, however, a purification artifact cannot be ruled out (Boman et al., 1985). Attacin-like factors have been described in *Manduca sexta*, *Drosophila melanogaster*, and *S. peregrina* (Spies et al., 1986, Flyg et al., 1987, and Ando et al., 1987).

Our initial experiments started out when we learned from the literature that the cecropins possessed a strong bactericidal activity against several bacteria including some animal pathogens (Steiner et al., 1981). Because of the peptide's broad activity it seemed reasonable to test its effect on plant pathogenic bacteria. Consequently, we synthesized cecropin B on a Biosearch Sam II peptide synthesizer and determined that it possessed substantial lytic activity against some of the most important plant pathogenic bacteria. Also, phytopathogenic fungi were tested and found to be sensitive to the peptide, these results are totally unique with no antecedents in the literature. In fact, it was stated that the activity of cecropins was limited to bacteria with no effect found against any eukaryote (Boman and Hultmark, 1987). These preliminary data provided us with the impetus to proceed further with the cloning and integration of the gene for cecropin B into plants. In order to obtain proper expression in the plant, it was necessary to introduce some modifications to the native sequence of the peptide. Therefore, we synthesized SB-37 and tested it against the same phytopathogens with similar results.

SB-37 differs from cecropin B in that two extra amino acids are added to the N-terminus of cecropin B (Met followed by Pro), Met^{11} substituted by a Val, and a C-terminal Gly. These N-terminal alterations do not destroy the bactericidal activity. This would seem contradictory to results obtained in other laboratories (Andreu et al., 1985). In addition, the presence of a C-terminal Gly, instead of the amidated Leu, has no affect on the lytic activity. Furthermore, significant changes in the amino acid sequence of cecropin B actually enhance biological activity as described recently by J. Jaynes and colleagues (Jaynes et al., 1988, 1989).

The "de novo" peptide, Shiva I (Fig. 7), designed to have significant differences in sequence homology, while conserving the overall charge distribution, amphipathy, and hydrophobic properties of the natural cecropin B has opened previously unexplored possibilities as new peptides in the Shiva series and in other classes of lytic peptides have been recently designed and tested with exciting results (manuscripts in prep.). In fact, the different types of peptides produced so far have proven to be more effective than cecropin B and SB-37 against several kinds of bacteria (Destéfano-Beltrán et al., 1990).

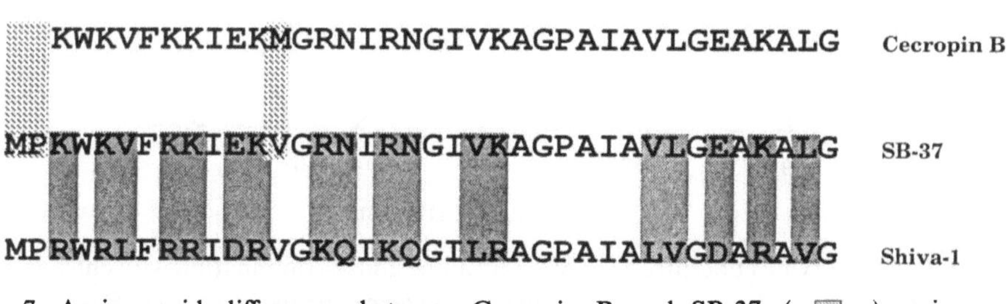

Fig. 7. Amino acid differences between Cecropin B and SB-37; ( ) amino acid differences between SB-37 and Shiva-1 ( )

An important characteristic of the humoral immunity of cecropia is that all these antibacterial proteins seem to act on different targets of the bacteria cells (Boman and Hultmark, 1987). This coordinated attack surely results in an overall enhanced activity as compared to that of the components alone. Therefore, it seemed important to test as much of the total system as possible. Due to the lack of cecropia lyzozyme and the attacins, we settled with the chicken egg-white lysozyme which is commercially available. Both peptides exhibited a pronounced synergistic effect in the presence of sublethal concentrations of the hen lysozyme (Fig. 8). This supports the notion that combining several antimicrobial genes into one cultivar will render the plant more resistant to pathogen attack.

Based on the in vitro data accumulated, we decided to proceed further with the construction of the DNAs. The assembling of the genes for SB-37 and Shiva-1 followed similar methodology. Six overlapping phosphorylated oligonucleotides for each construct were annealed and cloned into the BglII-EcoRI site of pMON530. This vector (kindly provided by S. Rogers, Monsanto Co.) carries a CaMV 35S promoter followed by a polylinker and a NOS 3′ polyadenylation signal. The constructs were then introduced into *A. tumefaciens* GV3111SE/pTiB6S3SE by triparental mating (Rogers et al., 1988). The integrity of the constructs were confirmed by reciprocal mating

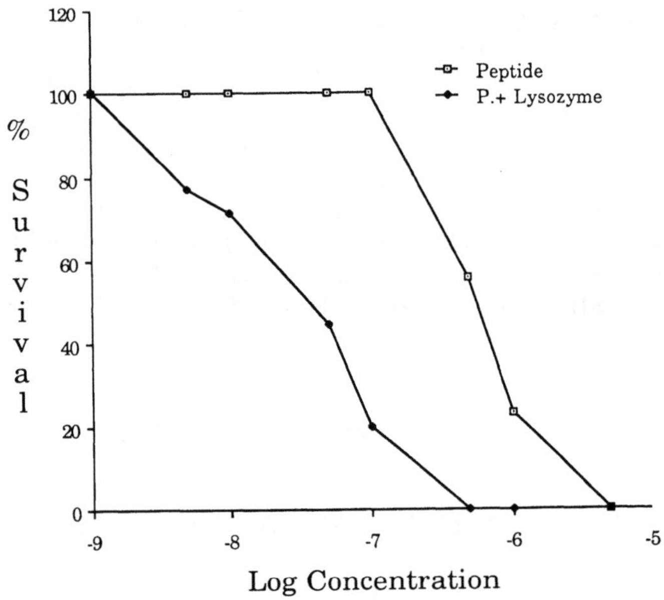

Fig. 8. Effects of varying concentrations of SB-37 alone and in combination with chicken egg-white lysozyme on in vitro survival of *Xanthomonas campestris* pv. *campestris*. Each point is the average of three replications

into *Escherichia coli* AJB361 (Dr. A. J. Biel, Louisiana State University). These and other "disease resistant" genes were subcloned into several other vectors (Destéfano-Beltrán et al., 1990). Additionally, in order to ensure adequate levels of expression, some of the constructs were placed under the control of the double 35S promoter (J. McPherson, University of British Columbia) and the chimeric fragments were introduced into the Hind III site of pBI 121 in the same direction of transcription as the NPTII and GUS genes. Wound-inducible modulation was obtained by placement of the genes under the control of the proteinase inhibitor II promoter from potato (L. Willmitzer, Berlin).

Tobacco was selected as a model system to test the efficacy of our constructs in enhancing plant disease resistance. Also, established tobacco transformation and regeneration protocols were available (Horsch et al., 1985). Most of our constructs have already been introduced into tobacco and F1 progeny are currently being evaluated for increased disease resistance (unpubl. data). Prior to selfing, all mother plants were screened for kanamycin resistance and GUS activity as appropriate and Southern analysis demonstrated the integration of single-copy genes. Northern and western analyses of some of the lines have shown the proper expression of the respective genes. In preliminary disease-challenge experiments, these plants exhibit delayed symptoms, reduced disease severity, and lower mortality after infection by *P. solanacearum* when compared to transgenic control plants (Fig. 9).

We have adopted a two-pronged approach to incorporate these genes into potato. Because of our previous experience with *A. rhizogenes* (Yang et al., 1989), we decided to start with our binary constructs, i.e., pBI

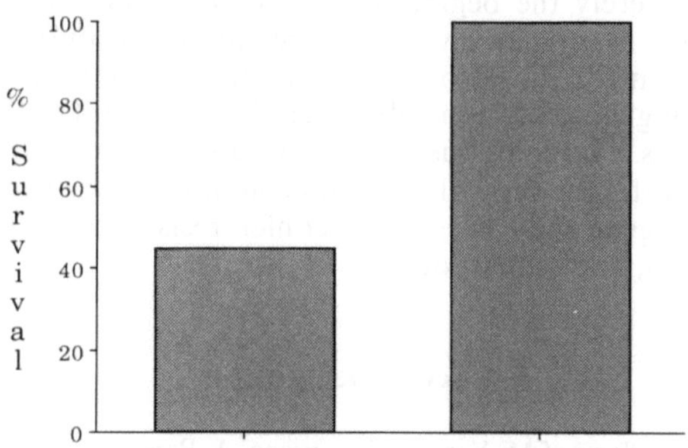

Fig. 9. Percent survival of tobacco transgenic plants expressing Shiva-1 under wound inducible control one month after stem-infection with *Pseudomonas solanacearum*

121-derivatives, in *A. rhizogenes* strain R1000 and infected stems of in vitro grown potato plants (Desiree, LT-9 and a yet to be named new cultivar from the Philippines, tentatively designated 86007). As expected, GUS-positive and kanamycin resistant hairy roots were obtained after two weeks of infection and plantlets were regenerated 3–4 weeks later. These plants have been evaluated and Southern analysis has confirmed the stable incorporation of these genes. The other approach utilized *A. tumefaciens* strain LBA 4404 containing the pBI 121-derivative constructs to infect leaf-disks of in vitro grown Desiree plants. GUS-positive and kanamycin resistant plants have been obtained after a few weeks. At present, most of the transformed lines are being propagated in order to conduct the pathogen challenge tests with different strains of *P. solanacearum* and *E. carotovora.*

IV. Conclusions

Although significant improvement of essential amino acid content in potato plant protein has not been achieved, we have developed a system which should, in time, allow for the improvement of the quality of plant proteins as we learn more about how to design proteins which are nutritionally complete.

In 1987, we first proposed that lytic peptide-encoding genes could be used to enhance overall resistance against bacterial and fungal diseases (Jaynes et al., 1987). Recently, it has been suggested that the transfer and expression of a new class of peptides found in honeybee, the apidaecins, could generate similar effects in plants (Casteels et al., 1989). However, these peptides lack lytic activity and are merely bacteriostatic, the significance of this difference still must be determined. Our promising results, so far obtained, are merely the beginning. The extreme malleability of peptide design portends many new avenues of research, not only in the arena of plant disease control, but also in human and veterinary medicine (Jaynes, 1989). Surprisingly, we have just discovered that several of the peptides kill plant nematodes. The consequences of this latest finding must be evaluated but it demonstrates the versatility of these biomolecules—perhaps, someday soon, a single gene may eventually confer resistance in plants to most bacterial, fungal, and nematode caused diseases.

Acknowledgements

We appreciate the efforts of M. Yang, V. Zambrano, A. Panta, C. Sigüeñas, R. Salinas, F. Buitrón, M. Juban, C. Clark, and T. Denny for various aspects of the work described in this article.

V. References

Agrios GN (1978) Plant pathology, 2nd edn. Academic Press, New York

Ando K, Okada M, Natori S (1987) Purification of sarcotoxin II, antibacterial proteins of *Sarcophaga peregrina* (flesh fly) larvae. Biochemistry 26: 226–230

Andreu D, Merrifield RB, Steiner H, Boman HG (1985) N-terminal analogues of cecropin A: Synthesis, antibacterial activity, and conformational properties. Biochemistry 24: 1683–88

Anonymous (1960) Index of plant diseases in the United States. Agriculture Handbook no 165. US Dept of Agriculture, Washington, DC

Boman HG, Hultmark D (1987) Cell-free immunity in insects. Annu Rev Microbiol 41: 103–26

Boman HG, Faye I, van Hofsten P, Kockum K, Lee J-Y, Xanthopoulos KG, Bennich H, Engstrøm Å, Merrifield RB, Andreu D (1985) On the primary structure of Lysozyme, Cecropins, and Attacins from *Hyalophora cecropia*. Dev Comp Immunol 9: 551–558

Casteels P, Ampe C, Jacobs F, Vaeck M, Tempst P (1989) Apidaecins: antibacterial peptides from honeybees. EMBO J 8: 2387–2391

Destéfano-Beltrán L, Nagpala P, Cetiner S, Dodds JH, Jaynes JM (1990) Enhancing bacterial fungal disease resistance in plants: application to potato. In: Vayda M, Park W (eds) The molecular and cellular biology of the potato. C.A.B. International, Wallindford, pp 205–221

Engstrøm Å, Xanthopoulos K, Boman HG, Bennich H (1985) Amino acid and cDNA sequences of lysozyme from *Hyalophora cecropia*. EMBO J 4: 2119–2122

Espinoza NO, Estrada R, Tovar P, Bryan J, Dodds JH (1984) Specialized potato technology, vol 1. International Potato Center, Lima, Perú, pp 1–20

Flyg C, Dalhammar G, Rasmuson B, Boman HG (1987) Insect immunity. Inducible antibacterial activity in Drosophila. Insect Biochem. 17: 153–160

Fraley RT, Rogers SG, Horsch RB, Kishore GM, Beachy R, Tumer N, Fischhoff DA, Delannay X, Klee HJ, Shah DM (1988) Genetic engineering for crop improvement. In: Gustafson JP, Appels R (eds) Chromosome structure and function. Plenum, New York, pp 283–298

Hansen R (1979) Food intake index. In: Hansen, RG, Wyse, BW, Sorenson AW (eds) Nutritional quality index of foods. AVI, Westport CT, pp 169–543

Horsch RB, Fry J, Hoffman N, Eichholtz D (1985) A simple and general method for transferring genes into plants. Science 223: 1229–1231

Hultmark D, Engstrøm Å, Andersson K, Steiner H, Bennich H, Boman HG (1983) Insect immunity: attacins, a family of antibacterial proteins from *Hyalophora cecropia*. EMBO J 2: 571–576

Hultmark D, Steiner H, Rasmusson T, Boman HG (1980) Purification and properties of three inducible bactericidal proteins from the hemolymph of immunized pupae of *Hyalophora cecropia*. Eur J Biochem 106: 7–16

Jaynes JM (1989) Lytic peptides portend an innovative age in the management and treatment of human disease. Drug News Perspect 3: 69–78

Jaynes JM, Juban M, Julian GJ, Miller MA (1990) Natural lytic peptides and their synthetic counterparts: a working structural model. (In preparation)

Jaynes JM, Julian GR, Jeffers GW, White KL, Enright FM (1989) In vitro cytocidal effect of lytic peptides on several transformed mammalian cell lines. Peptide Res 2: 157–160

Jaynes JM, Burton CA, Barr SB, Jeffers GW, Julian GR, White KL, Enright FM, Klei TR,

Laine RA (1988) In vitro cytocidal effect of novel lytic peptides on *Plasmodium falciparum* and *Trypanosoma cruzi.* FASEB J 2: 2878–2883

Jaynes JM, Xanthopoulos KG, Destéfano-Beltrán L, Doods JH (1987) Increasing bacterial disease resistance in plants utilizing antibacterial genes from insects. BioEssays 6: 263–270

Jaynes JM, Langridge P, Anderson K, Bond C, Sands D, Newman CW, Newman R (1985) Construction and expression of synthetic DNA fragments coding for polypeptides with elevated levels of essential amino acids. Appl Microbiol Biotech 21: 200–205

Newman CW, Jaynes JM, Stands DC (1980) Poly-L-lysine: a nutritional resource of lysine. Nutr Rep Int 22: 707–715

Okada M, Natori S (1983) Purification and characterization of an antibacterial protein from haemolymph of *Sarcophaga peregrina* (flesh fly) larvae. Biochem J 211: 727–734

Qu X, Steiner H, Engstrøm Å, Bennich H, Boman HG (1982) Insect immunity: isolation and structure of cecropins B and D from pupae of the Chinese oak silk moth, *Antheraea pernyi.* Eur J Biochem 127: 219–224

Rich AE (1983) Potato disease. Academic Press, New York

Rogers S, Klee H, Horsch RB, Fraley RT (1988) Use of cointegrating Ti plasmid vectors. In: Gelvin SB, Schilperoot RA (eds) Plant molecular biology manual. Kluwer, Dordrecht, pp A2: 1–12

Sawyer RL (1984) Potatoes for the developing world. International Potato Center, Lima, Peru

Shiba T, Ueki Y, Kubota I, Teshima T, Sugiyama Y, Oba Y, Kikuchi M (1984) Structure of lepidopteran, a self-defense substance produced by silkworm. In: Munekata E (ed) Peptide chemistry 1983. Protein Research Foundation, Osaka, pp 209–214

Spies AG, Karlinsey JE, Spence KD (1986) Antibacterial hemolymph proteins of *Manduca sexta.* Comp Biochem Physiol B 83: 125–133

Steiner H, Hultmark D, Engstrøm Å, Bennich H, Boman HG (1981) Sequence and specificity of two antibacterial proteins involved in insect immunity. Nature 292: 246–248

van Hofsten P, Faye I, Kockum K, Lee J-Y, Xanthopoulos KG, Boman IA, Boman HG, Engstrøm Å, Andreu D, Merrifield RB (1985) Molecular cloning, cDNA sequencing, and chemical synthesis of cecropin B from *Hyalophora cecropia.* Proc Natl Acad Sci USA 82: 2240–2243

Wellman FL (1972) Tropical american plant disease. Scarecrow, Metuchen, NJ

Xanthopoulos K, Lee JY, Gan R, Kockum K, Faye I, Boman HG (1988) The structure of the gene for cecropin B, an antibacterial immune protein from *Hyalophora cecropia.* Eur J Biochem 1: 371–376

Yang MS, Espinoza NO, Nagpala PG, Dodds JH, White FF, Schnorr KL, Jaynes JM (1989) Expression of a synthetic gene for improved protein quality in transformed potato plants. Plant Sci 64: 99–111

Chapter 3

Improvement of the Protein Quality of Seeds by Genetic Engineering

Mark A. Shotwell and Brian A. Larkins

Department of Plant Sciences, University of Arizona, Tucson, AZ 85721, U.S.A

With 8 Figures

Contents

I. Introduction

Seeds provide a major source of protein in the diets of humans and livestock. Accordingly, there has long been interest in the structure, amino acid content, and genetic regulation of seed proteins. Unfortunately, the value of seeds as protein sources is lessened by their unbalanced amino acid compositions. Most seed proteins are deficient in one or more of the amino acids that are essential for humans and other monogastric animals.

For many years, breeding programs have been directed at developing variants of crop plants that have a better balance of the essential amino acids (Nelson, 1969). These efforts have produced nutritionally improved varieties of many crops. However, the usefulness of these improved varieties as food sources is often compromised by undesirable traits such as lowered yield and increased susceptibility to pathogens (Nelson, 1980).

With the growth of the field of plant molecular biology has come increasing interest in improving the nutritional quality of crop plants by genetic engineering. Previous articles have considered possible strategies for crop improvement using recombinant DNA techniques (Larkins, 1982; Bright and Shewry, 1983). In this chapter we will summarize some recent results in which the promise of genetic engineering for the improvement of seed protein quality is beginning to be realized.

II. Background

During development, seeds accumulate large amounts of nitrogen in the form of protein. Most of this is a specific type called storage protein. The properties of storage proteins vary widely in different plants, but they have five characteristics in common: (1) they are synthesized only in developing seeds and in very large amounts; (2) they have no enzymatic activity; (3) they accumulate in aggregates called protein bodies; (4) they are often composed of a group of related polypeptides; and (5) they provide nitrogen, carbon, and sulfur to the germinating seedling. Two major types of storage proteins occur in seed plants: the globulins, which are found in the embryo and predominate in most dicot species, and the prolamines, which are only found in the endosperm of cereals. The globulins vary somewhat in solubility and amino acid composition, but their structures are conserved among seed plants (Casey et al., 1986). The alcohol-soluble prolamines have much more varied structures and are united only by their common hydrophobicity (Kreis and Tatham, 1990). In addition to the globulins and prolamines, other, less abundant, proteins are present in the seed. These components are often poorly characterized, but can significantly contribute to the overall protein quality of the seed.

Before any attempt can be made to genetically engineer seed proteins, the protein composition in the seed must be characterized. The genes encoding the proteins of interest must then be isolated, and their regulation analyzed. We will briefly describe some of the pertinent background work before considering some recent efforts to genetically engineer seed proteins.

A. Storage Globulins

Seed storage globulins are soluble in saline solutions, usually 0.4 to 1.0 M sodium chloride. They have been most intensively studied in the dicots, particularly the legumes, where they are stored primarily in the cotyledons (for reviews, see Casey et al., 1986; Shotwell and Larkins, 1989). In most monocot seeds, small amounts of globulin storage proteins are found in the embryo (scutellum). The globulins are rich in the amidated amino acids, asparagine and glutamine, consistent with their role as nitrogen stores. The primary nutritional deficiency of the storage globulins is a low level of the sulfur-containing amino acids, cysteine and methionine. Methionine is an essential amino acid for humans. Cysteine is synthesized from methionine, so that when methionine is limiting in the diet, cysteine also becomes essential.

Most of the storage globulins fall into two distinct structural groups with sedimentation coefficients of about 11S and 7S. The seeds of different species contain different relative amounts of the 11S and 7S types. Some legumes, like broad bean, store predominantly 11S globulins (Wright and Boulter, 1972), and others, like French bean, contain mostly 7S types (Derbyshire and Boulter, 1976). Structural similarities have been noted between the 11S and 7S storage globulins (Argos et al., 1985; Gibbs et al., 1989), and it has been proposed that all of the 11S and 7S globulins of flowering plants evolved from two ancestral proteins (Borroto and Dure, 1987).

The 11S storage globulins are isolated from seeds as hexameric complexes of 320 to 400 kDa, composed of six nonidentical subunits 52 to 65 kDa (Casey et al., 1986). Each subunit is made up of an acidic polypeptide (pI $\sim$6.5) of 33 to 42 kDa and a basic polypeptide (pI $\sim$9.0) of 19 to 23 kDa. The acidic and basic polypeptides are covalently linked by a single disulfide bond.

Physical measurements of the 11S storage globulins of legumes and other dicots have shown that in spite of their different molecular weights, they have very similar structures (Plietz and Damaschun, 1986). Within each subunit, the more hydrophobic basic polypeptide makes up the core of the molecule, and the more hydrophilic acidic polypeptide is arranged on the surface (Plietz et al., 1988). The quaternary structure consists of two trimers

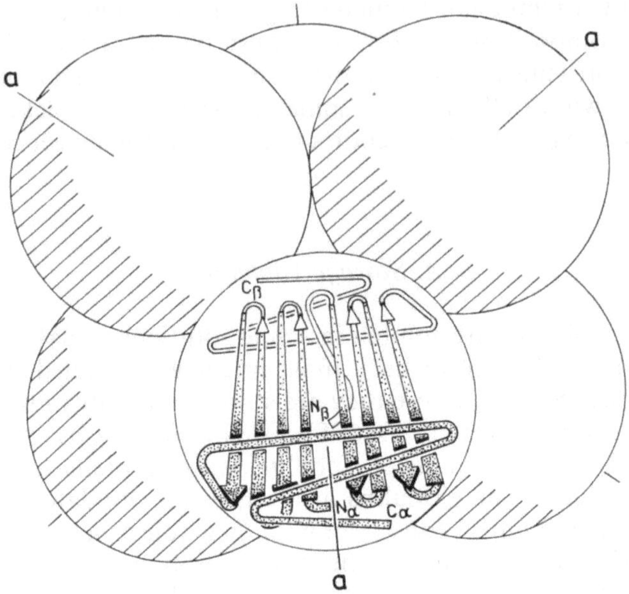

Fig. 1. Structural model for 11S storage globulins. The diagram is of a globulin hexamer consisting of two stacked trimers. The six axes of pseudo-symmetry (a). The polypeptide backbone of one subunit is shown. N_α and C_α, N and C termini of the acidic polypeptide; N_β and C_β, N and C termini of the basic polypeptide. Note that the C terminus of the basic chain is hydrophilic and forms the core of the subunit monomer and that the C terminus of the acidic chain (the hypervariable region) is hydrophilic and is on the surface of the molecule. From Plietz and Damaschun (1986) with permission of the authors

stacked one atop the other but offset by 60° (Fig. 1). Such a structure has been observed with the electron microscope (Reichelt et al., 1980).

Plietz et al. (1988) proposed a model in which the N-terminal regions of the acidic and basic polypeptides have a similar structure, consisting of alternating β-strands and β-turns (Fig. 1). The C terminus of the basic chain would be situated near the center of the molecule. Hydrophobic interactions between the C termini of adjacent basic polypeptides would stabilize the hexameric structure. In contrast, the C terminus of the acidic chain contains many polar and charged amino acid residues (Lycett et al., 1984; Nielsen et al., 1989). This hydrophilic domain, termed the hypervariable region, is predicted to have a helical conformation (Argos et al., 1985). This region would be exposed to the solvent at the surface of the molecule, where it would have little effect on the stability of the oligomer. The hypervariable region would thus be a good candidate for modifications to improve nutritional quality.

Compared to the 11S globulins, the 7S storage globulins have more varied structures. They are isolated from seeds as trimeric complexes of 145 to 190 kDa composed of three nonidentical subunits of 48 to 83 kDa (Casey et al., 1986). In some legumes, like pea, the subunit polypeptides undergo a series of proteolytic cleavages to yield polypeptides from 12 to 75 kDa (Spencer et al., 1983). In other species, including soybean (Meinke et al., 1981) and French bean (Brown et al., 1981), no proteolytic processing occurs. Most, if not all, 7S globulins completely lack cysteine, so these proteins are not linked by disulfide bonds.

The 7S globulins are co-translationally glycosylated in many legumes. For phaseolin of French bean, this first involves the transfer of an N-acetylglucosamine-mannose core oligosaccharide to one or two asparagine residues (Davies and Delmer, 1981; Bollini et al., 1983). Subsequently, several mannose units are removed from this high-mannose core and N-acetylglucosamine and xylose residues added during transport of phaseolin to the protein body (Herman et al., 1986). The transfer of the core oligosaccharide to nascent pea vicilin polypeptides is inhibited by tunicamycin (Badenoch-Jones et al., 1981). The inhibition of glycosylation does not affect the synthesis, assembly, or transport of the protein, however. Glycosylation of 7S globulins may thus not be an essential step in their pathway of synthesis and deposition.

Physical measurements have shown that the shape of the 7S globulins of legumes is approximated by a flattened, six-pointed spheroid (Plietz et al., 1983b). These spheroids consist of three nonidentical, Y-shaped subunits, each containing two rounded lobes of equal size. The three subunits are arranged around a three-fold symmetry axis and separated by deep solvent clefts (Plietz et al., 1983a). It is unclear what intermolecular interactions are involved in stabilizing the trimeric structure of the 7S globulins.

The 7S globulins contain small peptide inserts near the C terminus of the molecule (Argos et al., 1985). These peptides have a high content of acidic amino acids and are very hydrophilic. Like the longer hypervariable region of the 11S globulins, these peptides may be amenable to amino acid modification without affecting the three-dimensional structure of the molecule.

The 11S globulins are synthesized on rough endoplasmic reticulum (RER) and translocated into the lumen of the ER, where disulfide bond formation and assembly into 8S trimers occur (Barton et al., 1982; Chrispeels et al., 1982). These trimers are transported from the RER to the vacuole by way of Golgi-derived vesicles (Herman et al., 1976). After deposition in the vacuole, the globulin subunits are proteolytically cleaved into acidic and basic polypeptides, which remain linked by the disulfide bond. This cleavage reaction causes a conformational change that triggers the assembly of the 8S trimers into 11S hexamers (Dickinson et al., 1989). A similar intracellular pathway exists for the 7S globulins.

In legumes and other dicots, the globulin proteins are encoded by small multigene families. For the 11S globulins, estimates of gene numbers range from five for soybean glycinin (Nielsen et al., 1989) and sunflower helianthin (Vonder Haar et al., 1988) to eight for pea legumin (Domoney and Casey, 1985). The 7S types are encoded by somewhat more genes, for example seven for phaseolin (Talbot et al., 1984), 11 for pea vicilin (Domoney and Casey, 1985), and 15 to 20 for soybean β-conglycinin (Ladin et al., 1984). The globulin gene families in monocot species are larger still. There are about 50 genes encoding the 12S storage globulin of oats (Chesnut et al., 1989) and 15 to 25 genes for the structurally related storage glutelin of rice (Okita et al., 1989).

The structures of the storage globulin genes themselves are very well conserved among species. With only a few exceptions, genes encoding 11S globulins are interrupted by three introns whose positions in the coding sequence are maintained between species but whose lengths vary (Shotwell and Larkins, 1989). Fewer 7S globulin genes have been characterized, but to date all have been found to consist of six exons separated by five short introns (Doyle et al., 1986).

The degradation of storage globulins during seed germination appears to occur by a common pathway in legumes (Shutov and Vaintraub, 1987). Hydrolysis begins by the sequential action of cysteine proteinases that cleave internal peptide bonds. The first cysteine proteinase to act, designated proteinase A, generates sites for the action of a second type of cysteine proteinase, proteinase B, which is inactive against the native proteins. The C termini exposed by the concerted action of proteinases A and B are then attacked by carboxypeptidases, which are usually serine proteases. Finally, the tri- and dipeptide products of these enzymes are hydrolyzed into free amino acids by the action of amino- and dipeptidases.

B. Storage Prolamines

The prolamines have such varied structures that their only unifying characteristic is solubility in ethanol (in some cases the presence of a reducing agent) (Kreis and Tatham, 1990; Shotwell and Larkins, 1989). The prolamines are found only in monocots and are the predominant storage proteins in most of the agronomically important cereals. These proteins are rich in the amino acids proline and glutamine, but are deficient in the essential amino acids lysine, tryptophan, and threonine. The prolamines in members of the subfamily Festucoideae (wheat, rye, barley, oats, and rice) have some structural similarities, but differ considerably from those in the subfamily Panicoideae (maize, sorghum, and millet). The diversity of prolamine characteristics will be illustrated by a consideration of the prolamines of two important cereals, wheat and maize.

The wheat prolamines can be separated into three groups, the sulfur-rich gliadins, the sulfur-poor gliadins, and the high-molecular-weight (HMW) glutenins. Of these, the most abundant and diverse are the S-rich gliadins, which contain 2.5 to 3.5% cysteine plus methionine (Kreis et al., 1985b). They consist of the α-, β-, and γ-gliadins, which range in size from 32 to 44 kDa. The S-rich gliadins all contain extensive regions of repeated peptides rich in proline and glutamine. The S-poor prolamines of wheat ($<0.2\%$ Cys + Met) comprise the ω-gliadins, which are from 50 to 70 kDa and also contain extensive peptide repeats (Kreis et al., 1985a). Glutamine, proline, and phenylalanine make up 80% of the composition of the ω-gliadins. The third group of wheat prolamines, the HMW glutenins, range in size from 80 to 150 kDa and have an intermediate sulfur content. These proteins are extensively crosslinked by disulfide bonds and require dilute acid for solubility. The HMW glutenins also contain a region of peptide repeats, but these are unrelated to the ones in other wheat prolamines.

The prolamines of maize, the zeins, can be divided into four groups, the α-, β-, γ-, and δ-zeins, based on their solubility (Larkins et al., 1989). The α-zeins comprise a large number of polypeptides of 19 to 24 kDa that are heterogeneous in charge. The central region of these polypeptides consists of nine or ten tandemly repeated peptides of about 20 amino acid residues each. The β-zeins are polypeptides of 14 to 16 kDa that do not contain repetitive peptides and exhibit less charge heterogeneity than the α types. They also contain less proline and glutamine than the α-zeins, but more cysteine and methionine. The γ-zeins consist of two major components of 27 and 16 kDa and are made up of five distinct regions, including eight copies of the hexapeptide Pro-Pro-Pro-Val-His-Leu. These proteins are 25% proline and are also high in cysteine. The δ-zeins are a relatively minor fraction consisting of 10 kDa polypeptides that are extremely rich in the sulfur-containing amino acids.

A structural model has been proposed for the α-zeins of maize (Fig. 2) (Argos et al., 1982). In this model, the nine repeated peptides in the central region of the molecule are presumed to form hydrophobic α-helices separated by glutamine-rich turns. Hydrogen bonding between adjacent helices would align them into a cylindrical structure. The glutamine-rich turn regions would then form side chain interactions that would promote the packing of the cylindrical structures into dense aggregates in the protein bodies.

Prolamines are synthesized on RER and are deposited into protein bodies by two different mechanisms. In maize (Larkins and Hurkman, 1978), sorghum (Taylor et al., 1985), and rice (Krishnan et al., 1986), the prolamines aggregate into protein bodies directly in the RER. In wheat (Kim et al., 1988) and barley (Cameron-Mills and von Wettstein, 1980), on the other hand, at least some of the prolamines are transported from the ER to the vacuole by way of Golgi vesicles. It is in the vacuole that aggregation

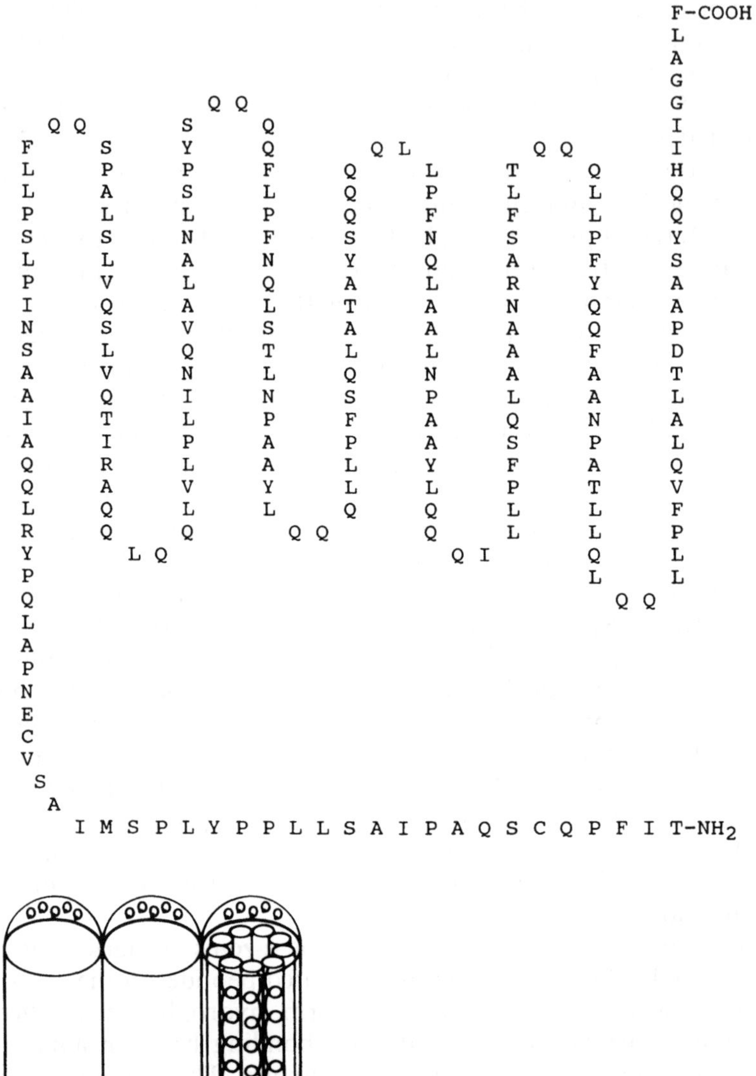

a

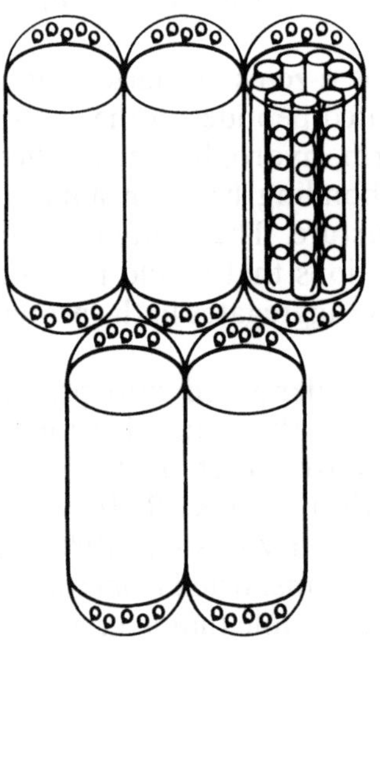

b

Fig. 2. Structure of the α-zeins of maize. A, Amino acid sequence of a 19 kDa α-zein. The nine peptide repeats in the central region of the molecule are aligned and are separated by glutamine-rich turns. B, Structural model for α-zeins. The peptide repeats from hydrophobic alpha helices that align into a cylinder (upper left). The glutamine-rich turns interact with adjacent molecules to allow tight packing of the molecules into dense aggregates. From Argos et al. (1982) with permission of the authors

of these prolamines into protein bodies occurs. By either pathway of deposition, there is no direct evidence of glycosylation of prolamines.

Prolamines may be deposited into proteins bodies in an ordered fashion. In maize, for example, the α-zeins have been shown to make up the core of the protein body and the β- and γ-zeins to be deposited around the periphery (Lending et al., 1988). Moreover, there is unequal distribution of the different zein types in different regions of maize endosperm. Just beneath the aleurone layer, the protein bodies contain large amounts of β- and γ-zeins, but nearer the center of the endosperm the protein bodies contain predominantly α-zeins. This pattern of zein deposition is thought to reflect changes in protein body structure that take place during endosperm development (Lending and Larkins, 1989).

The number of genes encoding prolamines varies widely for the different classes of proteins. For example, there are 75 to 100 genes encoding the α-zeins in maize (Wilson and Larkins, 1984), but only one or two genes each for the β-, γ-, and δ-zeins (Marks et al., 1984; Gallardo et al., 1988; Kirihara et al., 1988). The wheat α/β-glidins are encoded by a large family of greater than 100 genes (Okita et al., 1985), and the LMW glutenins by four subfamilies of two to twenty genes each (Colot et al., 1989). The prolamines of oats, the avenins, are encoded by about 25 gene copies (Chesnut et al., 1989), and the rice prolamines by 80 to 100 genes (Kim and Okita, 1988). Every prolamine gene isolated to date has been found not to contain introns.

Genes encoding prolamines differ from storage globulin genes in at least three ways. As described above, in many cases they are present in greater numbers in the genome than are the globulin genes. Second, the prolamine genes are not interrupted by introns as are the globulin genes. Third, sequence comparisons have shown that the prolamine and globulin genes have distinctly different codon usage patterns (Campbell and Gowri, 1990). These differences are primarily in the third base of the codons. Most prolamine genes have a much higher percentage of codons ending in C or G. Such a difference was recently found in oats, a cereal in which both globulin and prolamine genes have been characterized (Shotwell et al., 1990).

The pathway of degradation of storage prolamines during germination of cereal seeds is not as well characterized as that of the globulins of legumes. In several cereals the initial involvement of proteinase A has been observed, followed by the action of carboxypeptidases (Shutov and Vaintraub, 1987). In maize seeds, the degradation of zeins begins at about the second day of germination and is complete within about six days (Moureaux, 1979). Coincident with this period, an acidic endopeptidase can be detected that is active against undenatured zeins (Harvey and Oaks, 1974). The different classes of zeins are degraded sequentially. First to be hydrolyzed are the γ-zeins, which are found at the periphery of the protein bodies, followed by the α-zeins, which form the core of the protein body (Torrent et al., 1989). Recently, four closely related cysteine proteinases have been isolated from germinating maize seeds that may be involved in the first steps of zein degradation (de Barros and Larkins, 1990). These proteinases hydrolyze all four classes of zeins, both when they are dissolved in alcohol and when they are aggregated in protein bodies.

III. Factors To Be Considered Before Modifying Seed Proteins

The characterization of storage proteins from various species has clearly shown that these proteins have very ordered structures and pathways of synthesis and deposition. It is therefore unlikely that random or gross modifications of their primary amino sequences can be made without negative consequences to their synthesis and accumulation within the seed. Accordingly, several factors must be considered before modifying the genes encoding seed storage proteins in vitro and re-introducing them into plants.

A. Solubility Characteristics

The amino acid sequence of the modified seed proteins should not be altered in a way that significantly changes their solubility. For example, the introduction of a large number of polar and charged residues into prolamines would increase their solubility in aqueous solution. Such an increased solubility might interfere with the aggregation of the prolamines within the ER and the formation of protein bodies.

B. Secondary Structures

The introduction of amino acids into storage proteins could interfere with the formation of secondary structures that are a prerequisite of aggregation. For example, adding cysteines to 7S globulins could lead to the formation of disulfide bonds that would disrupt its normal conformation. In the prolamines, the insertion or replacement of amino acids in their repeat regions might prevent hydrogen bond formation or other necessary interactions between adjacent repeats. This would interfere with the tight packing of the polypeptides, which may be required for deposition in the protein bodies.

C. Proteolytic Cleavages

The cleavage of the 11S globulin precursors into acidic and basic polypeptides appears to cause a conformational change that catalyzes the assembly of 8S trimers into 11S hexamers (Dickinson et al., 1989). If the site of this cleavage reaction is altered, processing of the precursors might be prevented and assembly might not occur. It would thus appear that this region of the 11S globulins should not be modified. No physiological significance has been found for the proteolytic processing of the 7S globulins, however. It is likely that their sequences could be changed in a way that prevents processing without affecting assembly and deposition.

D. Interactions Between Polypeptides

Specific interactions must occur between the acidic and basic polypeptides of the 11S globulins for assembly into trimers and then into hexamers to take place. Modifications of the basic chain have been shown to perturb its structure in such a way as to prevent trimer formation (C.D. Dickinson et al., unpubl. results). The basic polypeptide of the 11S globulins would thus not be a good candidate for extensive modifications.

E. Assembly of Oligomers

Assembly of storage globulin subunits into trimers and hexamers may be a required step in the pathway of deposition of these proteins. Recent studies have suggested that one type of 11S globulin subunit of soybean (glycinin G2) is incapable of assembling into 8S trimers in the absence of other subunits (C.D. Dickinson et al., unpubl. results). This subunit would thus be a poor choice for expression by itself in transgenic plants. Whether individual 7S subunits are similarly incapable of assembly into trimers is unknown.

F. Intracellular Transport

Molecular signals that specify the route of transport of storage proteins from the ER to the protein body have yet to be elucidated. It appears that glycosylation does not play a major role in intracellular targeting, however. The targeting signal for phytohemagglutinin in French bean was found to be contained in its polypeptide sequence and not in its polysaccharide component (Voelker et al., 1989). Whatever the nature of the targeting signals in storage proteins, it is clear that they must be maintained if the modified proteins are to be deposited in protein bodies. Alternatively, targeting signals must be introduced into proteins that are to be expressed in transgenic plants.

G. Protein Stability

The modified proteins must be stable after their synthesis in seeds of transgenic plants and not be subject to proteolytic degradation. Modifications that interfere with the normal assembly and aggregation of storage proteins or that do not maintain correct intracellular targeting signals are likely to render the protein susceptible to degradation. Protein instability may especially prove to be an obstacle to the expression of certain monocot proteins in dicots and vice versa.

H. Gene Numbers

As described earlier, some storage proteins, notably certain prolamines, are encoded by large multigene families. It is not clear how many of the genes in the multigene families are transcriptionally active. Nevertheless, engineered genes would presumably need to be expressed at a high rate compared to the native genes to ensure that enough of the modified protein is synthesized to affect the overall protein composition of the seed. If an engineered α-zein gene is re-introduced into corn, for example, it must be under the transcriptional control of an extremely active promoter, since potentially as many as 100 unmodified α-zein genes are active in endosperm tissue.

I. Codon Distribution

There is evidence that certain foreign genes introduced into host organisms are not highly expressed because their codon distributions are unlike that of highly expressed genes in the host organism (deBoer and Kastelein, 1986). For these genes to be actively expressed, their codon usage patterns may first have to be altered to more closely resemble the preferred pattern in host genes. This may require changing much of the coding sequence to the synonymous codons preferred by the host organism.

J. Tissue Specificity and Developmental Regulation

Engineered genes must have promoters that direct transcription only in seeds and at the correct stages of seed development. These considerations are especially relevant when attempting to express a monocot gene in a dicot host or vice versa. For example, in one study an α-zein gene was found to be expressed at low levels in leaves, stems, and flowers of transgenic petunia plants when controlled by its own promoter sequence (Ueng et al., 1988).

K. *Degradation After Germination*

Finally, the modified protein should preferably not be altered in such a way that it is no longer a substrate for the enzymes that degrade storage proteins after seed germination. Seeds containing large amounts of proteins resistant to degradation might not germinate efficiently. Since the proteinases that initiate storage protein degradation and the carboxypeptidases that act secondarily seem not to be sequence-specific (Shutov and Vaintraub, 1987), large alterations in storage proteins should be tolerated.

IV. Strategies for Improving Seed Proteins

A. *In vitro Mutagenesis of Seed Storage Protein Genes to Increase Limiting Amino Acids*

1. Prolamines

As described above, the prolamine storage proteins are deficient in the essential amino acids lysine and tryptophan. To create a nutritionally superior zein protein, Wallace et al. (1988) introduced lysine and tryptophan codons into an α-zein cDNA sequence. Single lysine replacements were made in the N-terminal coding sequence as well as within and between the peptide repeats (Fig. 3). These single-replacement constructs were recombined to generate double-replacement constructs. In addition, short oligonucleotides encoding lysine- and tryptophan-rich peptides were inserted at several points in the coding sequence.

Methods are not yet available for transforming DNA sequences into maize cells and regenerating plants. Therefore, Wallace et al. (1988) used a heterologous system to evaluate the effects of these modification on the synthesis and aggregation of zeins. Synthetic mRNAs were transcribed in vitro and injected into *Xenopus laevis* oocytes. To detect the modified zein proteins, ^{3}H-leucine was injected into the oocytes 24 h after mRNA injection. After a 4 h labelling period, the oocytes were homogenized and the zein-containing protein bodies separated by centrifugation in metrizamide gradients.

Wallace et al. (1988) first injected total native zein mRNAs into oocytes (Fig. 4A), as well as synthetic mRNA encoding unmodified α-zein (Fig. 4B). This showed that zein mRNAs are translated in *Xenopus* oocytes and that the zein polypeptides assemble into protein bodies that sediment deep in metrizamide gradients. They then found that the synthesis of α-zein protein in *Xenopus* oocytes was unaffected by the addition of a single lysine residue (Fig. 4D), two lysine residues (Fig. 4E), or short lysine- and tryptophan-rich

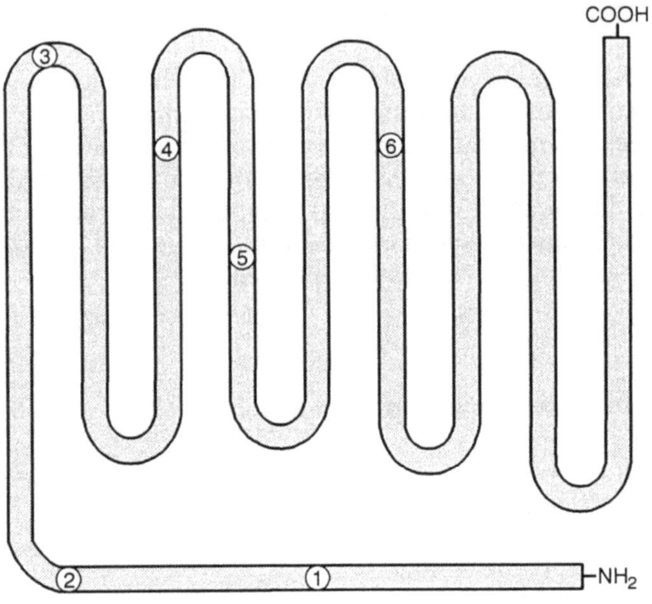

MODIFICATIONS

(1a) Ile → Lys

(1b) Insert of AlaThrTrpLysThrTrpLysThr

(2) Insert of SV40 VP2

(3) Gln → Lys

(4) Asn → Lys

(5) Insert of LysThrTrpLysThr

(6) Asn → Lys

Fig. 3. Modifications of an α-zein. Shown is a schematic representation of a 19 kDa α-zein derived from the amino acid sequence in Fig. 2A. 1–6, relative locations of seven amino acid substitutions and insertions. For a complete description of all of the modifications that were made to the molecule, see Wallace et al. (1988)

oligopeptides (Fig. 4F, G). Moreover, protein bodies containing the modified zeins sedimented deep in the gradients, indicating that they had densities similar to those containing unmodified α-zeins (Fig. 4B). These results showed that the introduction of lysine and tryptophan into α-zeins does not affect their aggregation into protein bodies. As a control, the coding sequence for a 17 kDa fragment of the VP2 protein of SV40 was inserted into the zein sequence. This large alteration did not interfere with

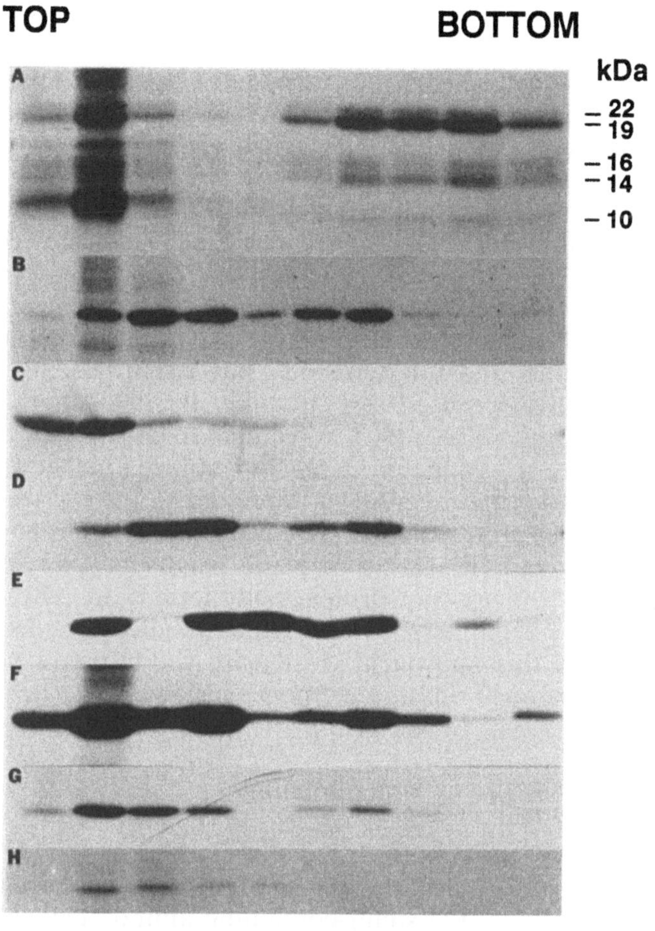

Fig. 4. Density gradient separation of zein-containing protein bodies from oocytes injected with zein mRNAs. *Xenopus laevis* oocytes were injected with synthetic zein mRNAs and ^{3}H-leucine as described by Wallace et al. (1988). The oocytes were then homogenized and fractionated by centrifugation in gradients of 10–15% metrizamide. The distribution of radioactivity in the gradients was determined by polyacrylamide gel electrophoresis and fluorography. Panels correspond to the following mRNAs (see Fig. 3 for the locations of the modifications): A, total native zein mRNAs; B, unmodified synthetic α-zein mRNA; C, a control in which exogenous ^{3}H-labelled α-zein was added to the oocyte homogenate; D, isoleucine to lysine substitution at position 1; E, asparagine to lysine double substitution at positions 4 and 6; F, oligopeptide insertion at position 1; G, oligopeptide insertion at position 5; H, insertion of a 17 kDa fragment of the VP2 protein of SV40 at position 2. The relative masses of the zeins are indicated to the right of A. The zeins in B–G migrated at 19 kDa; the modified zein in H migrated at 35 kDa. TOP and BOTTOM, relative position in the gradient. Adapted from Wallace et al. (1988)

zein synthesis, but did prevent zein aggregation into protein bodies (Fig. 4H).

The synthesis of lysine-containing α-zeins has also been analyzed in the seeds of transgenic tobacco plants. Preliminary studies showed that α-zein promoter sequences do not function efficiently in solanaceous plants (Ueng et al., 1988). Accordingly, Williamson et al. (1988) placed the zein coding sequences under the transcriptional control of a promoter from a gene for phaseolin, the major 7S globulin of French bean. These chimeric constructs were introduced into tobacco by *Agrobacterium tumefaciens*-mediated transformation.

High levels of mRNA transcripts were produced from these chimeric constructs in petunia and tobacco seeds, but α-zein proteins were barely detectable (Williamson et al., 1988; Ohtani et al., 1990). Both normal and lysine-containing zein polypeptides were found to be unstable, disappearing from the seed with a half-life of about four hours. Furthermore, the zein proteins were observed as aggregates appressed to the cell wall in petunia seeds rather than in protein bodies (Wallace et al., 1990). It is possible that the instability of α-zeins in transgenic plants results from a requirement for β-, γ-, and possibly δ-zeins for protein body formation. Alternatively, the accumulation of α-zeins may be deleterious to dicot cotyledonary cells. Continued use of this approach to increase the nutritive value of zein proteins awaits a transformation method for maize.

2. Globulins

Modifications have been made to genes encoding both the 7S and 11S storage globulins of legumes in order to overcome their primary nutritive deficiency, low levels of the sulfur-containing amino acids. In one recent study, Hoffman et al. (1988) modified a gene for β-phaseolin to increase its methionine content from three to nine residues. This was accomplished by the insertion of a 45-bp oligonucleotide containing six methionine codons into the third exon of a β-phaseolin genomic clone (Fig. 5). They then introduced both this high-methionine gene and a normal β-phaseolin gene into tobacco by *Agrobacterium*-mediated transformation and examined the synthesis and deposition of phaseolin in seeds of regenerated transgenic plants.

Hoffman et al. (1988) detected similar levels of phaseolin mRNAs in seeds of plants transformed with the normal and high-methionine genes and found that the expression of the two genes was both seed-specific and developmentally regulated. In addition, the normal and high-methionine phaseolin polypeptides were both glycosylated at two sites and correctly assembled into trimers. However, the accumulation of the high-methionine phaseolin in tobacco seeds was, on average, only about 0.2% of that of the

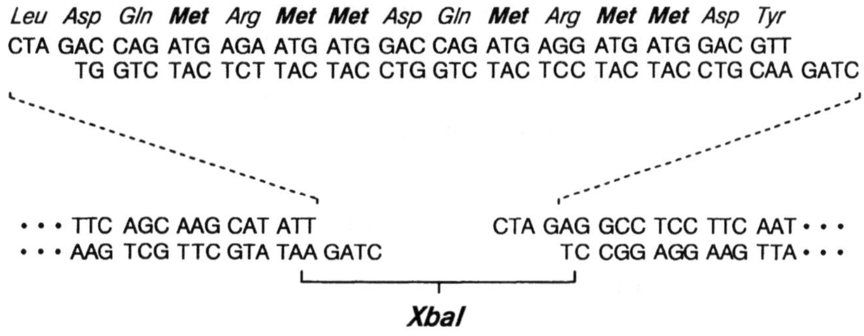

Fig. 5. Sequence of a methionine-encoding oligonucleotide inserted into a phaseolin gene. The oligonucleotide (top) was inserted into a unique XbaI site in the third exon of a phaseolin genomic sequence (bottom). The addition of these 45 nucleotides increased the methionine content of phaseolin from three to nine. This modified gene was introduced into tobacco, and the accumulation of high-methionine phaseolin in seeds was monitored. Adapted from Hoffman et al. (1988)

normal protein (Table 1). Electron microscopy revealed that modified phaseolin polypeptides were present in the ER and Golgi vesicles in transformed seeds, but not in protein bodies.

The results of Hoffman et al. (1988) indicated that the high-methionine phaseolin was synthesized in the ER, assembled into trimers, and transported to the Golgi apparatus in transgenic tobacco seeds like normal phaseolin in developing bean cotyledons. Unlike normal phaseolin, however, the modified proteins were then degraded, either in the Golgi vesicles or perhaps in the protein bodies. The site in the phaseolin polypeptide into which the methionine-rich peptide was inserted has recently been identified as being important for stabilizing the phaseolin trimers (Lawrence et al., 1990). The rapid turnover of the high-methionine phaseolin may have thus resulted from a distortion of the secondary structure in this region rendering the trimers less stable and more susceptible to proteolytic degradation.

The results of this and the previously described study demonstrate that modifications of seed storage proteins to increase their nutritive value may in some cases be accompanied by changes in protein structure that interfere with intracellular transport and stability. Such structural changes must be avoided if this approach is to yield nutritionally improved seeds.

In a somewhat different approach, Dickinson et al. (1990) developed in vitro synthesis and assembly assays that allow the evaluation of storage protein modifications without having to introduce the engineered storage protein genes into plants. Dickinson et al. (1990) first introduced a series of insertions and deletions into a cDNA sequence encoding glycinin, the major 11S storage globulin of soybean. They then used SP6 RNA polymerase to

Table 1. Amount of phaseolin detected by enzyme-linked immunosorbent assay in seeds of tobacco plants transformed with unmodified and high-methionine phaseolin genes

Plant	Phaseolin (ng/ml)
Unmodified	
P.1	2010
P.5	1890
P.7	11,040
P.8	6960
Average	5475
High-methionine	
hiP.6	10.8
hiP.21	5.0
hiP.23	5.4
hiP.27	18.5
Average	10.0

Data from Hoffman et al. (1988)

transcribe the modified cDNAs in vitro and translated the transcripts in rabbit reticulocyte lysates to produce radiolabelled glycinin subunits. The ability of the modified glycinins to assemble into 9S trimers either alone (self-assembly) or in the presence of unmodified glycinin (mixed assembly) was determined as illustrated in Fig. 6.

Using these assays, Dickinson et al. (1990) found that the assembly of glycinin subunits into trimers was less sensitive to alterations in the acidic polypeptide than in the basic polypeptide. Deletions at the C terminus of the acidic chain, the hypervariable region, had very little effect on assembly. These results are consistent with structural models of 11S globulins in which the basic polypeptide contributes more to the structure of the subunit than does the acidic chain and the hypervariable region resides on the surface and has little to do with determining structure (Plietz et al., 1988).

Some of the modifications made by Dickinson et al. (1990) were designed to introduce methionine into glycinin subunit polypeptides. In one series of modifications, they inserted one, three, or five copies of the oligonucleotide 5′-CGCATG-3′ into a unique restriction site in the coding region for the hypervariable region. These insertions had the effect of adding one, three, or five copies of the dipeptide Arg-Met. All of these modified

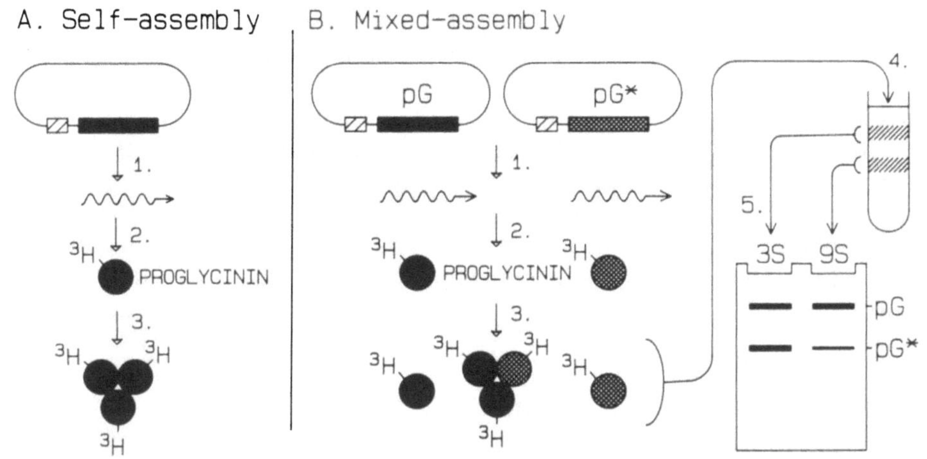

Fig. 6. Diagram of in vitro synthesis and assembly assays for soybean glycinins. A, Self-assembly assay. Glycinin cDNA sequences are modified by the introduction of short oligonucleotide insertions and deletions and transcribed in vitro using SP6 RNA polymerase (1). The modified glycinin mRNAs are then translated in a rabbit reticulocyte lysate in the presence of ^{3}H-leucine (2). The ^{3}H-labelled proteins are then allowed to assemble into oligomers during a 30 h incubation at 25 °C. B, Mixed assembly assay. Normal (pG) and modified (pG*) glycinin cDNA sequences are transcribed in the same reaction (1), and the mixture of mRNAs is translated as described above to yield ^{3}H-labelled proteins (2). The mixed proteins are then allowed to assemble during a 12 h incubation at 25 °C. In both assays, the proteins are fractionated by centrifugation in sucrose gradients into 3S monomers and 9S trimers (4), which are then separated by polyacrylamide gel electrophoresis and visualized by fluorography (5). From Dickinson et al. (1990) with permission of the authors and publisher

subunits self-assembled into trimers efficiently, and the one modified subunit tested in the mixed assembly assay assembled almost as rapidly as unmodified glycinin.

This work is an important contribution in two respects. First, it provides additional information on the structural requirements for assembly of 11S storage globulins, which seems to be a prerequisite of deposition in protein bodies. The results clearly demonstrate that the hypervariable region is the best candidate for insertion of methionine-containing peptides. Second, the assembly assays developed in these studies allow the effects of storage protein modifications to be evaluated in vitro in a very short time. This permits the rapid identification of the most promising modified proteins, which may then be analyzed in transgenic plants, a much more time-consuming procedure.

B. Introduction of Heterologous Genes Encoding Proteins That Contain Limiting Amino Acids

An alternative strategy for improving the nutritive value of seeds is to introduce a protein not normally found in the seeds of a given plant that has an amino acid composition that will compensate for deficiencies in that plant. Since the amino acid inadequacies of the storage globulins and prolamines are different, a more balanced amino acid composition could be achieved by the synthesis of globulins in monocot seeds or by the synthesis of prolamines in dicot seeds.

1. Prolamines in Dicots

In one recent study, Hoffman et al. (1987) introduced a gene encoding the sulfur-rich β-zein of maize into tobacco, a dicot. β-Zeins contain about 7% methionine and 4% cysteine (Pedersen et al., 1986), the two most critically lacking amino acids in dicot seed protein. To increase the likelihood that the maize gene would be expressed in tobacco, they replaced its 5′ and 3′ flanking sequences with analogous regions of a gene encoding β-phaseolin. This chimeric gene was introduced into the tobacco genome by *Agrobacterium*-mediated transformation, and transgenic plants were regenerated and analyzed for β-zein synthesis.

Hoffman et al. (1987) found that the accumulation of β-zein polypeptides varied by a factor of 80 in the 19 transgenic plants they examined, which contained from one to eight gene copies. The greatest accumulation was in a plant containing seven genes, in which β-zein made up 1.6% of the total seed protein. The synthesis of the β-zein in transgenic tobacco was restricted to seeds.

The subcellular site of deposition of the β-zein in tobacco seeds was determined using immunogold labelling techniques. Most of the β-zein labelling was observed within the protein bodies, primarily in the crystalloid component (Fig. 7). Considerably more β-zein was found in the protein bodies in embryo tissue than in endosperm. Thus, although zeins normally aggregate directly in the ER of maize endosperm, in transgenic tobacco seeds β-zein is targeted to vacuolar protein bodies in both endosperm and embryo tissue. The pattern of synthesis of β-zeins in tobacco that Hoffman et al. (1987) observed contrasts with that seen for α-zeins. As described earlier, α-zein polypeptides were synthesized in tobacco seeds, but they were not targeted into protein bodies and were rapidly degraded (Ohtani et al., 1990).

Hoffman et al. (1987) did not evaluate the effect of β-zein accumulation on the amino acid composition of tobacco seeds. It is likely, however, that even the relatively modest amounts of synthesis of β-zein polypeptides

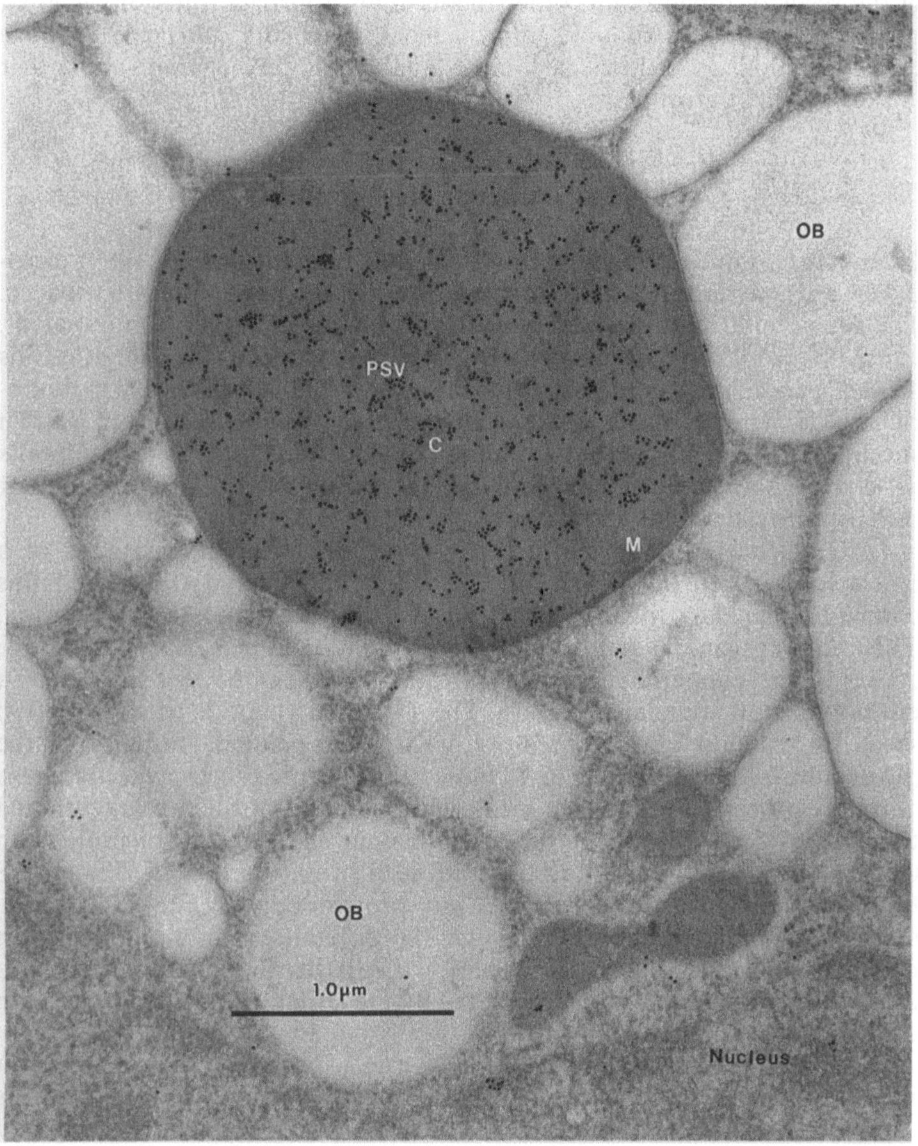

Fig. 7. Localization of β-zein polypeptides in protein bodies of transgenic tobacco seeds. The micrograph is of a portion of a tobacco embryo cell and shows one protein storage vacuole (PSV) or protein body. The majority of the immunogold labelling of β-zeins occurred within the crystalloid region (C). Much less labelling took place within the matrix of the protein body (M). There was no labelling above background in the oil bodies (OB) and nucleus. × 37,400. From Hoffman et al. (1987) with permission of the authors and publisher

measured in this study would significantly increase the methionine content of the seeds of tobacco and other dicots. This approach thus shows great promise for the nutritional improvement of dicots, particularly once transformation procedures become available for the legumes and other important dicot crop plants.

2. 2S Albumins in Dicots

In another recent example of this strategy, Altenbach et al. (1989) introduced a gene encoding the 2S storage protein of the Brazil nut into tobacco. The Brazil nut 2S protein is very rich in the sulfur-containing amino acids, consisting of 18% methionine and 8% cysteine (Ampe et al., 1986). It is thus an ideal candidate to compensate for the amino acid deficiencies of globulin-rich seeds. The 2S protein is a water-soluble albumin made up of two subunits of 9 and 3 kDa that are linked by disulfide bonds. These subunits arise from the three-step proteolytic processing of a 17 kDa precursor polypeptide (Sun et al., 1987).

Altenbach et al. (1989) attached the 5′ and 3′ flanking regions of a phaseolin gene to the Brazil nut 2S coding region and transferred this chimeric gene into tobacco by *Agrobacterium*-mediated transformation. The four transgenic plants they examined contained between one and five copies of the chimeric gene, and all four plants expressed the gene in a developmentally regulated pattern. The 9 kDa polypeptide was detected in seeds by reaction of protein blots with a monoclonal antibody to the purified Brazil nut subunit (Fig. 8), indicating that the 17 kDa precursor was correctly processed in tobacco seeds. The Brazil nut protein was estimated to constitute up to 8% of the seed protein that was extractable with 0.25 M NaCl. After separating the total seed proteins into 11S, 7S, and 2S fractions, they detected the Brazil nut protein only in the 2S fraction, suggesting that normal assembly of the subunits into monomers was occurring in tobacco. The 2S protein was rapidly degraded after germination of tobacco seeds; no traces of the 9 kDa subunit remained by the 5th day of germination.

Synthesis of the Brazil nut 2S protein in tobacco caused a significant increase in the methionine content of the seeds. In the four transgenic plants examined, the increases in methionine were 24, 32, 9.7, and 20% compared to an untransformed control plant (Table 2). The 32% increase in plant number 32 corresponds to an 8% content of the Brazil nut 2S protein in the seeds of that plant. Contrary to expectations, there was no increase in the cysteine content of the transgenic seeds (Table 2), although the Brazil nut 2S protein contains about 8% cysteine. It was also found that the proportion of 2S proteins was not increased in transgenic seeds relative to 11S and 7S proteins. Altenbach et al. (1989) concluded that concomitant with the

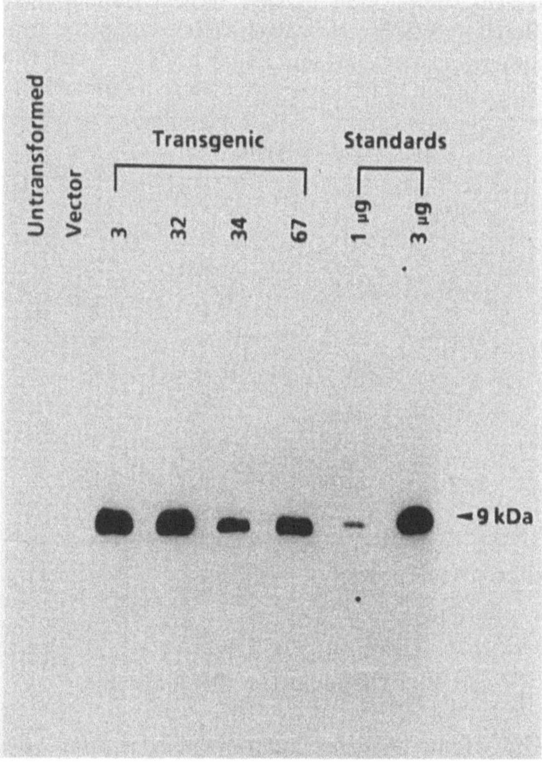

Fig. 8. Detection of Brazil nut 2S protein in seeds of transgenic tobacco. Total salt-extractable proteins from mature tobacco seeds were analyzed by immunoblotting using a monoclonal antibody to the 9 kDa subunit of the Brazil nut 2S protein. 3, 32, 34, and 67, contain 30 µg of seed proteins from plants transformed with the chimeric 2S protein gene. The first two lanes on the left contain protein samples from control plants, either untransformed or transformed with only the vector plasmid. The two lanes on the right are standards containing 1 and 3 µg of the 2S protein purified from Brazil nuts. From Altenbach et al. (1989) with permission of the authors and publisher

synthesis of the Brazil nut 2S protein in tobacco, there was a decrease in the synthesis of some tobacco 2S proteins. These tobacco proteins would presumably contain high levels of cysteine and low levels of methionine.

This study demonstrates that a useful alternative approach to compensate for the nutritive deficiencies of seeds is to introduce genes encoding foreign proteins with desirable characteristics rather than to modify the naturally occurring seed protein genes. It is likely that this strategy will be frequently employed once transformation techniques become available for legumes and cereals.

Table 2. Amino acid composition of seed protein from four tobacco plants expressing the Brazil nut 2S protein chimeric gene compared to untransformed control tobacco. For simplicity, only four amino acids are listed. Data are given as mole percent (and percentage change from control)

		Transformed			
Amino acid	Control	3	32	43	67
Ser	5.30	5.18 (−3)	5.30 (0)	5.64 (+6)	5.35 (+1)
Glx	19.65	19.56 (0)	19.86 (+1)	19.56 (0)	19.77 (+1)
Met	3.60	4.47 (+24)	4.74 (+32)	3.95 (+10)	4.31 (+20)
1/2 Cys	2.23	2.09 (−6)	2.12 (−5)	2.27 (+2)	2.21 (−1)

Data from Altenbach et al. (1989)

V. Prospects for the Future

Achieving the goal of developing nutritionally improved crops using the techniques of molecular biology depends upon continued progress in at least three research areas. First, in addition to further characterization of the globulin and prolamine seed storage proteins, we need a greater understanding of the other protein components of the seed. An analysis of these non-globulin, non-prolamine proteins may reveal the existence of proteins in low abundance that have desirable nutritive characteristics. For example, there may be low-abundance, non-prolamine proteins in cereal seeds that contain moderate to high levels of lysine. An engineered increase in the synthesis of these minor components would increase the overall protein quality of cereal seeds. Knowledge of minor seed proteins and the genes that encode them is a prerequisite of such a strategy.

As an alternative to modifying major seeds storage proteins, increasing the synthesis of minor seed proteins, or introducing genes encoding foreign seed proteins, it has now become practical to chemically synthesize genes encoding nutritionally complete proteins. Recently, a synthetic gene coding for a polypeptide rich in the essential amino acids was introduced into potato (Yang et al., 1989; cf. Destéfano-Beltrán et al., 1991). Once we gain sufficient understanding of the structural requirements for seed protein synthesis, assembly, transport, and stability, this technique may have wide application in the engineering of improved seeds.

Second, we require a more detailed understanding of the molecular events that cause high levels of synthesis of storage proteins in developing seeds. Efforts are underway in several plant systems to identify the *cis*-acting DNA sequences and *trans*-acting protein factors that together determine the tissue-specificity and developmental timing of storage protein gene expression. A thorough comprehension of the function these components will provide the basis for increasing the synthesis of desirable seed proteins and, equally useful, decreasing the synthesis of undesirable seed proteins. In addition to transcriptional regulation, we need to know more about the role of post-transcriptional mechanisms in storage protein synthesis. Codon usage patterns, mRNA structure and stability, and amino acid biosynthesis may all influence the expression of genetically engineered seed storage proteins.

Finally, and most critically, further progress depends upon the development of transformation and regeneration systems for the agronomically important legumes and cereals. Some progress has recently been reported toward these goals, but many technical hurdles remain. It is essential that methods be devised, whether by biological or mechanical means, for the stable introduction of altered seed protein genes into legumes and cereals with the relative ease of *Agrobacterium*-mediated transformation of solanaceous species. These methods must then be coupled with regeneration procedures that yield whole plants expressing the altered protein.

VI. References

Ampe C, Van Damme J, de Castro LAB, Sampaio MJAM, Van Montagu M, Vandekerckhove J (1986) The amino-acid sequence of the 2S sulphur-rich proteins from seeds of Brazil nut (*Bertholletia excelsa* H.B.K.). Eur J Biochem 159: 597–604

Altenbach SB, Pearson KW, Meeker G, Staraci LC, Sun SSM (1989) Enhancement of the methionine content of seed proteins by the expression of a chimeric gene encoding a methionine-rich protein in transgenic plants. Plant Mol Biol 13: 513–522

Argos P, Pederson K, Marks MD, Larkins BA (1982) A structural model for maize zein proteins. J Biol Chem 257: 9984–9990

Argos P, Narayana SVL, Nielsen NC (1985) Structural similarity between legumin and vicilin storage proteins from legumes. EMBO J 4: 1111–1117

Badenoch-Jones J, Spencer D, Higgins TJV, Millerd A (1981) The role of glycosylation in storage-protein synthesis in developing pea seeds. Planta 153: 201–209

Barton KA, Thompson JF, Madison JT, Rosenthal R, Larvis NP, Beachy RN (1982) The biosynthesis and processing of high molecular weight precursors of soybean glycinin subunits. J Biol Chem 257: 6089–6095

Bollini R, Vitale A, Chrispeels MJ (1983) In vivo and in vitro processing of seed reserve protein in the endoplasmic reticulum: evidence for two glycosylation steps. J Cell Biol 96: 999–1007

Borroto K, Dure L III (1987) The globulin seed storage proteins of flowering plants are derived from two ancestral genes. Plant Mol Biol 8: 113–131

Bright SWJ, Shewry PR (1983) Improvement of protein quality in cereals. CRC Crit Rev Plant Sci 1: 49–92

Brown JWS, Bliss FA, Hall TC (1981) Linkage relationships between genes controlling seed proteins in French bean. Theor Appl Genet 60: 251–259

Cameron-Mills V, von Wettstein D (1980) Protein body formation in the developing barley endosperm. Carlsberg Res Commun 45: 577–594

Campbell WH, Gowri G (1990) Codon usage in higher plants, green algae, and cyanobacteria. Plant Physiol 92: 1–11

Casey R, Domoney C, Ellis N (1986) Legume storage proteins and their genes. Oxford Surv Plant Mol Cell Biol 3: 1–95

Chesnut RS, Shotwell MA, Boyer SK, Larkins BA (1989) Analysis of avenin proteins and the expression of their mRNAs in developing oat seeds. Plant Cell 1: 913–924

Chrispeels MJ, Higgins TJV, Spencer D (1982) Assembly of storage protein oligomers in the endoplasmic reticulum and processing of the polypeptides in the protein bodies of developing pea cotyledons. J Cell Biol 93: 306–313

Colot V, Bartels D, Thompson R, Flavell R (1989) Molecular characterization of an active wheat LMW glutenin gene and its relation to other wheat and barley prolamin genes. Mol Gen Genet 216: 81–90

Davies HM, Delmer DP (1981) Two kinds of protein glycosylation in a cell-free preparation from developing cotyledons of *Phaseolus vulgaris*. Plant Physiol 68: 284–291

de Barros EG, Larkins BA (1990) Purification and characterization of zein-degrading proteinases from germinating maize endosperm. Plant Physiol (in press)

de Boer HA, Kastelein RA (1986) Biased codon usage: an exploration of its role in optimization of translation. In: Reznikoff W, Gold L (eds) Maximizing gene expression. Butterworths, Boston, pp 225–285

Derbyshire E, Boulter D (1976) Isolation of legumin-like protein from *Phaseolus aureus* and *Phaseolus vulgaris*. Phytochemistry 15: 411–414

Destéfano-Beltrán L, Nagpala P, Jaeho K, Dodds JH, Jaynes JM (1991) Genetic transformation of potato to enhance nutritional value and confer disease resistance. In: Dennis ES, Llewellyn DJ (eds) Molecular approaches to crop improvements. Springer, Wien New York, pp 17–32 [Dennis ES et al (eds) Plant gene research. Basic knowledge and application]

Dickinson CD, Hussein EHA, Nielsen NC (1989) Role of post-translational cleavage in glycinin assembly. Plant Cell 1: 459–469

Dickinson CD, Scott MP, Hussein EHA, Argos P, Nielsen NC (1990) Effect of structural modifications on the assembly of a glycinin subunit. Plant Cell 2: 403–413

Domoney C, Casey R (1985) Measurement of gene number for seed storage proteins in *Pisum*. Nucleic Acids Res 13: 687–699

Doyle JJ, Schuler MA, Godette WD, Zenger V, Beachy RN, Slightom JL (1986) The glycosylated seed storage proteins of *Glycine max* and *Phaseolus vulgaris*. J Biol Chem 261: 9228–9238

Gallardo D, Reina M, Rigau J, Boronat A, Palau J (1988) Genomic organization of the 28 kDa glutelin-2 gene from maize. Plant Sci 54: 211–218

Gibbs PEM, Strongin KB, McPherson A (1989) Evolution of legume seed storage proteins — a domain common to legumins and vicilins is duplicated in vicilins. Mol Biol Evol 6: 614–623

Harvey BMR, Oaks A (1974) The hydrolysis of endosperm protein in *Zea mays*. Plant Physiol 53: 453–457

Herman EM, Shannon LM, Chrispeels MJ (1986) The Golgi apparatus mediates the transport and post-translational modification of protein body proteins. In: Shannon LM, Chrispeels MJ (eds) Molecular biology of seed storage proteins and lectins. American Society of Plant Physiologists, Rockville, MD, pp 163–173

Hoffman LM, Donaldson DD, Bookland R, Rashka K, Herman EM (1987) Synthesis and protein body deposition of maize 15-kd zein in transgenic tobacco seeds. EMBO J 6: 3213–3221

Hoffman LM, Donaldson DD, Herman EM (1988) A modified storage protein is synthesized, processed, and degraded in the seeds of transgenic plants. Plant Mol Biol 11: 717–729

Kim WT, Okita TW (1988) Structure, expression, and heterogeneity of the rice seed prolamines. Plant Physiol 88: 649–655

Kim WT, Franceschi VR, Krishnan HB, Okita TW (1988) Formation of wheat protein bodies: involvement of the Golgi apparatus in gliadin transport. Planta 176: 173–182

Kirihara JA, Hunsperger JP, Mahoney WC, Messing JW (1988) Differential expression of a gene for a methionine-rich storage protein in maize. Mol Gen Genet 211: 477–484

Kreis M, Tatham AS (1990) The prolamin storage proteins of cereal seeds: structure and evolution. Biochem J 267: 1–12

Kreis M, Forde BG, Rahman S, Miflin BJ, Shewry PR (1985a) Molecular evolution of the seed storage proteins of barley, rye and wheat. J Mol Biol 183: 499–502

Kreis M, Shewry PR, Forde BG, Forde J, Miflin BJ (1985b) Structure and evolution of seed storage proteins and their genes with particular reference to those of wheat, barley and rye. Oxford Surv Plant Mol Cell Biol 2: 253–317

Krishnan HB, Franceschi VR, Okita TW (1986) Immunochemical studies on the role of the Golgi complex in protein-body formation in rice seeds. Planta 169: 471–480

Ladin BF, Doyle JJ, Beachy RN (1984) Molecular characterization of a deletion mutation affecting the a′-subunit of β-conglycinin of soybean. J Mol Appl Genet 2: 372–380

Larkins BA (1982) Genetic engineering of seed storage proteins. In: Kosuge T, Meredith CP, Hollaender A (eds) Genetic engineering of plants. An agricultural perspective. Plenum, New York, pp 93–118

Larkins BA, Hurkman WJ (1978) Synthesis and deposition of zein in protein bodies of maize endosperm. Plant Physiol 62: 256–263

Larkins BA, Lending CR, Wallace JC, Galili G, Kawata EE, Geetha KB, Kriz AL, Martin DM, Bracker CE (1989) Zein gene expression during maize endosperm development. In: Goldberg RB (ed) The molecular basis of plant development. Alan R Liss, New York, pp 109–120

Lawrence MC, Suzuki E, Varghese JN, Davis PC, Van Donkelaar A, Tulloch PA, Colman PM (1990) The three-dimensional structure of the seed storage protein phaseolin at 3 Å resolution. EMBO J 9: 9–15

Lending CR, Larkins BA (1989) Changes in the zein composition of protein bodies during maize endosperm development. Plant Cell 1: 1011–1023

Lending CR, Kriz AK, Larkins BA, Bracker CE (1988) Structure of maize protein bodies and immunocytochemical localization of zeins. Protoplasma 143: 51–62

Lycett GW, Delauney AJ, Zhao W, Gatehouse JA, Croy RRD, Boulter D (1984) Two cDNA clones coding for the legumin protein of *Pisum sativum* L. contain sequence repeats. Plant Mol Biol 3: 91–96

Marks MD, Pedersen K, Wilson DR, DiFonzo N, Larkins BL (1984) Molecular biology of the maize seed storage proteins. Curr Top Plant Biochem Physiol 3: 9–18

Meinke DW, Chen J, Beachy RN (1981) Expression of storage-protein genes during soybean seed development. Planta 153: 130–139

Moureaux T (1979) Protein breakdown and protease properties of germinating maize endosperm. Phytochemistry 18: 1113–1117

Nelson OE (1969) Genetic modification of protein quality in plants. Adv Agron 21: 171–194

Nelson OE (1980) Genetic control of polysaccharide and storage protein synthesis in the endosperms of barley, maize, and sorghum. In: Pomeranz Y (ed) Advances in cereal science and technology. American Association of Cereal Chemistry, St. Paul, pp 41–71

Nielsen NC, Dickinson CD, Cho T-J, Thanh VH, Scallon BJ, Fischer RL, Sims TL, Drews GN, Goldberg RB (1989) Characterization of the glycinin gene family in soybean. Plant Cell 1: 313–328

Ohtani T, Wallace JC, Thompson GA, Galili G, Larkins BA (1990) Normal and lysine-containing zeins are unstable in transgenic tobacco seeds. Plant Mol Biol (in press)

Okita TW, Cheesbrough V, Reeves CD (1985) Evolution and heterogeneity of the α-/β-type and γ-type gliadin DNA sequences. J Biol Chem 260: 8203–8213

Okita TW, Hwang YS, Hnilo J, Kim WT, Aryan AP, Larson R, Krishnan HB (1989) Structure and expression of the rice glutelin multigene family. J Biol Chem 264: 12573–12581

Pedersen K, Argos P, Naravana SVL, Larkins BA (1986) Sequence analysis and characterization of a maize gene encoding a high-sulfur zein protein of M_r 15,000. J Biol Chem 261: 6279–6284

Plietz P, Damaschun G (1986) The structure of the 11S seed globulins from various plant species: comparative investigations by physical methods. Stud Biophys 3: 153–173

Plietz P, Damaschun G, Müller JJ, Schlesier B (1983a) Comparison of the structure of the 7S globulin from *Phaseolus vulgaris* in solution with the crystal structure of 7S globulin from *Canavalia ensiformis* by small angle X-ray scattering. FEBS Lett 162: 43–46

Plietz P, Damaschun G, Zirwer D, Gast K, Schlesier B (1983b) Structure of 7S seed globulin from *Phaseolus vulgaris* L. in solution. Int J Biol Macromol 5: 356–360

Plietz P, Drescher B, Damaschun G (1988) Structure and evolution of the 11S globulins: conclusions from comparative evaluation of amino acids sequences and X-ray scattering data. Biochem Physiol Pflanzen 183: 199–203

Reichelt R, Schwenke K-D, König T, Pähtz W, Wangermann G (1980) Electron microscopic studies for estimation of the quaternary structure of the 11S globulin (helianthin) from sunflower seed (*Helianthus annuus* L.). Biochem Physiol Pflanzen 175: 653–663

Shotwell MA, Larkins BA (1989) The biochemistry and molecular biology of seed storage proteins. In: Marcus A (ed) The biochemistry of plants. A comprehensive treatise, vol 15. Academic Press, San Diego, pp 297–345

Shotwell MA, Boyer SK, Chesnut RS, Larkins BA (1990) Analysis of seed storage protein genes of oats. J Biol Chem 265: 9652–9658

Shutov AD, Vaintraub IA (1987) Degradation of storage proteins in germinating seeds. Phytochemistry 23: 75–94

Spencer D, Chandler PM, Higgins TJV, Inglis AS, Rubira M (1983) Sequence interrelationships of the subunits of vicilin from pea seeds. Plant Mol Biol 2: 259–267

Sun SSM, Altenbach SB, Leung FW (1987) Properties, biosynthesis and processing of a sulfur-rich protein in Brazil nut (*Bertholletia excelsa* H.B.K.). Eur J Biochem 162: 477–483

Talbot DR, Adang MJ, Slightom JL, Hall TC (1984) Size and organization of a multigene family encoding phaseolin, the major seed storage protein in *Phaseolus vulgaris* L. Mol Gen Genet 198: 42–49

Taylor JRN, Schüssler L, Liebenberg, NvdW (1985) Protein body formation in starchy endosperm of developing *Sorghum bicolor* (L.) Moench seeds. S Afr J Bot 51: 35–40

Torrent M, Geli MI, Ludevid MD (1989) Storage-protein hydrolysis and protein-body breakdown in germinated *Zea mays* L. seeds. Planta 180: 90–95

Ueng P, Galili G, Sapanara V, Goldsbrough PB, Dube P, Beachy RN, Larkins BA (1988) Expression of a maize storage protein gene in petunia plants is not restricted to seeds. Plant Physiol 86: 1281–1285

Voelker TA, Herman EM, Chrispeels MJ (1989) In vitro mutated phytohemagglutinin genes expressed in tobacco seeds: role of glycans in protein targeting and stability. Plant Cell 1: 95–104

Vonder Haar RA, Allen RA, Cohen EA, Nessler CL, Thomas TL (1988) Organization of the sunflower 11S storage protein gene family. Gene 74: 433–443

Wallace JC, Galili G, Kawata EE, Cuellar RE, Shotwell MA, Larkins BA (1988) Aggregation of lysine-containing zeins into protein bodies in *Xenopus* oocytes. Science 240: 662–664

Wallace JC, Ohtani T, Lending CR, Lopes M, Williamson JD, Shaw KL, Gelvin SB, Larkins BA (1990) Factors affecting physical and structural properties of maize protein bodies. In: Lamb C, Beachy RN (eds) Plant gene transfer. Alan R Liss, New York [UCLA symposium on molecular and cellular biology, new series, vol 129] (in press)

Williamson JD, Galili G, Larkins BA, Gelvin SB (1988) The synthesis of a 19 kilodalton zein protein in transgenic *Petunia* plants. Plant Physiol 88: 1002–1007

Wilson DM, Larkins BA (1984) Zein gene organization in maize and related grasses. J Mol Evol 29: 330–340

Wright DJ, Boulter D (1972) The characterization of vicilin during seed development in *Vicia faba* (L.). Planta 105: 60–65

Yang MS, Espinoza NO, Nagpala PG, Dodds JH, White FF, Schnorr KL, Jaynes JM (1989) Expression of a synthetic gene for improved protein quality in transformed potato plants. Plant Sci 64: 99–111

Chapter 4

Novel Insect Resistance Using Protease Inhibitor Genes

Angharad M.R. Gatehouse, Donald Boulter, and Vaughan A. Hilder

Department of Biological Sciences, University of Durham,
South Road, Durham DH1 3LE, U.K.

With 3 Figures

Contents

I. Introduction

Crop plants have been primarily selected for high yields, nutritional value, including low mammalian toxicity and, where necessary, adaptation to certain environmental conditions. This "selection pressure" over the centuries has severely disrupted the co-evolutionary relationships between plants and insects with the consequence that very few cultivated species have retained the degree of resistance exhibited by their wild relatives (Feeny, 1976). For example, since the complex phenolic, gossypol, is toxic to mammals and so interferes with the utilisation of cotton seed meal as an animal feed, lines devoid of this compound have been selected for. This has resulted in the new improved lines being extremely susceptible to attack by the cotton budworm *Heliothis virescens* (Berardi and Goldblatt, 1980) towards which gossypol is toxic. An additional problem is encountered when crop plants are introduced into foreign geographical regions and so are exposed to an array of pests and diseases to which they have not had the opportunity to evolve any defence mechanisms. A classic example is illustrated by potato damage. On its introduction from Bolivia into the

South Western United States the crop became exposed to the Colorado beetle (*Leptinotarsa decemlineata*) towards which it had no inherent resistance; this insect pest has subsequently become established as the most serious insect pest of potato worldwide. In 1988 the insecticide expenditure for protection of potato was estimated to be US$ 198 million (Table 1).

It has been estimated that approximately 37% of all crop production is lost to pests and diseases with 13% lost to insects. These unacceptable losses occur despite the exceptionally high annual expenditure on pesticides. In an attempt to redress the balance, and particularly as a consequence of the increased concern about the heavy reliance on chemical insecticides in recent years, some breeders and biotechnologists are now attempting to exploit inherent resistance either by conventional plant breeding (Redden et al., 1983; Harmsen et al., 1988) or by recombinant DNA technology (Hilder et al., 1987).

The most obvious advantages of breeding for resistance, as opposed to an almost exclusive reliance upon chemical pesticides are:

(*i*) it provides season long protection,
(*ii*) insects are always treated at the most sensitive stage,
(*iii*) protection is independent of the weather,
(*iv*) it protects plant tissues which are difficult to treat using insecticides. For example chemical insecticides are inadequate at controlling larvae of *Ceutorhyncus assimilis* (pest of oil seed rape) since they attack the developing ovules within the pods,
(*v*) only crop-eating insects are exposed,
(*vi*) the material is confined to the plant tissues expressing it and therefore do not leach into the environment,
(*vii*) the active factor is biodegradable and choice of suitable genes/gene products can ensure it is not toxic to man and animals.

Another major advantage of producing resistant crop plants is that there are considerable financial savings. In 1988 it was estimated that the cost of chemical insecticide protection for cotton, maize and rice alone was US$ 3808 million (Table 1). Pest resistant crops would thus offer economic advantages to the farmer which should ultimately benefit the consumer.

Genetic manipulation has further potential advantages over conventional plant breeding in that it enables desired gene(s) to be transferred to the recipient plants without the co-transfer of undesirable characteristics and also allows the transfer of genes across incompatibility barriers. There are, however, constraints involved with this new technology, e.g., the inability to transform and regenerate specific crop cultivars, availability of 'useful' genes and Governmental regulatory barriers.

To date two different strategies for producing transgenic insect resistant plants have been used. One uses toxic proteins of bacterial origin, whilst the other utilises insecticidal proteins of plant origin. Although it is not the

Table 1. Insecticide expenditure by territory of four major crops

Crop	Major territory	Insecticide expenditure[a]	Application expenditure[b]
Cotton	USA	305	200
	USSR	275	53
	China	180	24
	India	175	23
	Other	595	86
Rice	Japan	450	165
	China	148	20
	India	128	17
	Other	319	45
Maize	USA	235	35
	Europe, USSR	120	40
	Other	145	25
Potato	Europe	80	—
	Far East	58	—
	USA	40	—
	Other	20	—

Values given are × US$ 10^{-6}
[a] Nat West 1988 expenditure
[b] AGC Ltd. (estimates) 1987 expenditure

purpose of this chapter to consider the use of bacterial toxins in detail nor to compare the efficacy of one strategy with the other, as this has recently been reviewed in detail (Gatehouse et al., 1990; Hilder et al., 1989) the use of such compounds will be considered briefly.

II. Bacterial Toxins

The bacterial toxin with insecticidal properties that has received most attention to date is the crystalline protein produced by strains of *Bacillus thuringiensis* (Bt). The idea of using such compounds is nothing new since Bt toxins in various formulations (usually as whole sporulated bacteria) have been employed as insecticidal crop sprays for more than 20 years (Dulmage, 1981), but their use is limited by high costs of production. Furthermore the toxic components, i.e., intracellular crystals which are produced by the bacteria during sporulation, are not stable under field spraying and exposure conditions, particularly under high uv light. When ingested by certain insects these crystals are typically partially hydrolysed under the alkaline conditions of the midgut so releasing proteins of M_r 65,000–160,000; further proteolytic processing of these proteins releases smaller toxic fragments. The mechanism of toxicity appears to depend on the disruption of the membranes of cells lining the gut and this action is only carried out by the cleaved fragments (Sacchi et al., 1986). These Bt toxins are very species-specific and although commercial toxins effective against more than 50 lepidopteran pest species have been identified, comparatively few with activity against coleopterans have been isolated. Despite this possible limitation, the potential for increasing the utility of Bt by the genetic engineering of plants to express the Bt toxin was recognised at an early stage and in 1987 transgenic tobacco and tomato plants expressing the Bt toxin at levels insecticidal to certain lepidopteran insects, notably *Manduca sexta*, were obtained (Vaeck et al., 1987; Fischhoff et al., 1987; Barton et al., 1987).

III. Protease Inhibitors

Plant protease inhibitors vary in size from M_r 4000–80,000 (Richardson, 1977) but are usually in the range of 8000–20,000; this is particularly true of the legume serine protease inhibitors. Many of these inhibitors, especially those that have a subunit molecular weight around 10,000 are capable of simultaneously inhibiting two molecules of enzyme per molecule of inhibitor. This is thought to have arisen by gene duplication–elongation (Odani and Ikenaka, 1972) and these particular ones are known as 'double-headed' or Bowman-Birk type inhibitors. However, the two enzymes which the inhibitor molecule inhibits need not necessarily be the same; for example the Bowman-Birk inhibitor from soybean simultaneously inhibits one molecule of trypsin and one of chymotrypsin (Seidl and Liener, 1971). This is also true for one of the double-headed inhibitors isolated from cowpea (Gatehouse et al., 1980; Gennis and Cantor, 1976); however the majority of Bowman-Birk type isoinhibitors from cowpea were found to only inhibit trypsin. Not only do some of these double-headed inhibitors inhibit different enzymes of the same type but some have been described which inhibit two different classes of enzyme (Shivaraj and Pattabiraman, 1981; Campos and Richardson, 1983). Association with their respective proteases is pH dependent, being strong at neutral pH and rapidly decreasing as the pH is lowered from neutrality towards pH 3.0. The mechanism of action of these inhibitors appears to be competitive in that the inhibitor molecule and substrate molecule share the same enzyme binding site (Laskowski and Sealock, 1971).

Potato and tomato plants contain two powerful inhibitors of serine proteases called Inhibitor I (monomer M_r 8100) and Inhibitor II (monomer

M_r 12,300) (Plunkett et al., 1982). Inhibitor I is an inhibitor of chymotrypsin that only weakly inhibits trypsin at its single reactive site whilst inhibitor II contains two reactive sites, one of which inhibits trypsin while the other inhibits chymotrypsin. Both inhibitors are synthesised as precursors and undergo post-translational modification (Cleveland et al., 1987) to form the mature proteins which are sequestered into the vacuole (Shumway et al., 1970).

IV. Protease Inhibitors as Protective Agents

Interest in the effects of plant protease inhibitors towards insect attack was aroused as early as 1947 when Mickel and Standish observed that larvae of certain pests were unable to develop normally on soybean products (Mickel and Standish, 1947). These observations led Lipke and co-workers (Lipke et al., 1954) to study the toxicity of soybean inhibitors on the development of *Tribolium confusum*, a common pest of stored grain. Although negative with respect to soybean trypsin inhibitors, these studies did reveal the presence of a specific inhibitor of *Tribolium* larval digestive proteolysis; this was later isolated and shown to completely inhibit larval gut proteolysis of both *T. confusum* and *T. castaneum* (Birk et al., 1963) but was inactive towards either mammalian trypsin or chymotrypsin (Applebaum and Konijn, 1966). Subsequently the protease inhibitors from the legume *Vigna unguiculata* were also shown to be antimetabolic to developing *T. confusum* larvae when tested in artificial diet (Gatehouse and Hilder, 1988). Applebaum et al. (1964) demonstrated that lima bean inhibitor, ovomucoid, soybean Kunitz inhibitor and the soybean Bowman-Birk inhibitor inhibited one of the midgut proteases isolated from larvae of *Tenebrio molitar* which is known to contain both trypsin and chymotrypsin-like enzymes (Zwilling, 1968), whereas the inhibitor of the *Tribolium* digestive protease, on the other hand, was found to be ineffective. The Kunitz inhibitor was subsequently demonstrated to inhibit larval growth and delay pupation of young corn borer larvae, *Ostrinia nubilalis*, when incorporated into the diet at levels of 2–5% (Steffens et al., 1978). However, since maize inhibitors were found to have no effect upon growth and metamorphosis of these larvae when added to the diet at levels of 2–3% the authors suggested that the proteolytic enzyme inhibitors of legumes may be related to the resistance of these plants to the corn borer.

Perhaps one of the most convincing pieces of evidence for the direct involvement of protease inhibitors as protective agents against insects was in 1972 when Green and Ryan demonstrated that wounding of the leaves of potato or tomato plants by adult Colorado beetles, or their larvae, induced a rapid accumulation of two inhibitors throughout the aerial tissues of the plants (Green and Ryan, 1972); this effect of insect damage could be simulated by mechanical wounding of the leaves. A factor, or wound hormone, called proteinase inhibitor-inducing factor (PIIF), was found to be released from the damaged leaves and transported to other leaves where it initiated synthesis and accumulation of inhibitors. Within two to three days after the attack; the two inhibitors could account for over 1% of the soluble proteins of the leaf, and they remained in the leaves for long periods of time, stored in the central vacuoles (Shumway et al., 1976; Walker-Simmons and Ryan, 1977). This finding strongly implicates protease inhibitors in plant protection (Ryan, 1983). Recent evidence suggests that they are one part of the complex interaction between plant nutritional value and the insects' digestive physiology (Broadway et al., 1986).

The first demonstration of a protective role for protease inhibitors in 'field' resistance was in 1979 when Gatehouse et al. found that the levels of protease inhibitors (CpTI) present in seeds of a resistant line of cowpea, TVu 2027, was correlated with resistance to a major insect pest, *Callosobruchus maculatus*. The resistant variety of cowpea contained significantly higher levels of inhibitors, at least twice as much as any other variety and in the order of three to four times more than the majority of the varieties tested. Differences in the inhibitor content of the resistant variety compared to susceptible varieties were shown to be quantitative and not qualitative (Gatehouse et al., 1979). The antimetabolic properties of the purified cowpea trypsin inhibitor were demonstrated in feeding trials with the larvae of *C. maculatus*, a pest species (Gatehouse et al., 1979). Initial feeding trials using protein fractions extracted from the resistant cowpea line showed that the albumin fraction (containing >95% of the protease inhibitors) was toxic, whereas the globulin fraction was not. Removal of trypsin inhibitors from the albumin fraction rendered it non-toxic. To confirm the toxicity of the trypsin inhibitors further feeding trials were carried out by adding the purified inhibitor to an artificial diet at a range of concentrations. Since the differences in trypsin inhibitors between different cowpea lines was considered to be primarily in the amounts of inhibitors present, rather than the resistant one containing a unique "effective" inhibitor, trypsin inhibitor was purified from commercially available non-resistant cowpea lines, rather than from the resistant line TVu 2027. Addition of inhibitor to diet at a level of 0.8%, which is marginally lower than the physiological concentration found in the resistant seeds, resulted in complete larval mortality. These results confirm that the trypsin inhibitors play a major role in conferring seed resistance in this particular example of 'field' resistance. However, subsequent studies on the mechanism of inheritance of bruchid beetle resistance suggested that resistance was not solely attributed to protease inhibitors but in fact was very complex (Redden et al., 1983); this was recently confirmed by Xavier-Filho et al. (1989).

V. Engineering for Resistance Using Cowpea Protease Inhibitor Genes

The first successful example of genetic engineering of plants for insect resistance using genes of plant origin was achieved using a cowpea protease inhibitor (CpTI) gene (Hilder et al., 1987). These inhibitors form ideal candidates for genetic engineering for insect resistance since they have been demonstrated to be insecticidal towards a wide range of economically important field and storage pests, including members of the Lepidoptera such as *Heliothis* sp. and *Spodoptera* sp., Coleoptera such as *Diabrotica* sp., *Bruchidae* sp. and *Anthonomus* sp., and Orthoptera such as *Locusta* sp. when tested in artificial diets (Table 2). Furthermore, and of prime importance, they appear to exhibit low or no mammalian toxicity (unpubl. data). These protease inhibitors are small polypeptides of about 80 amino acids and belong to the Bowman-Birk type of double-headed serine protease inhibitors and whilst most bind two molecules of trypsin, some of the isoinhibitors are trypsin/chymotrypsin inhibitors (Gatehouse et al., 1980). These inhibitors are products of a middle repetitive gene family.

The CpTI gene used to produce transgenic plants encoded a trypsin/trypsin inhibitor and was derived from plasmid pUSSR c3/2, a member of a complementary DNA library prepared from cowpea cotyledon polyadenylated RNA (Fig. 1) (Hilder et al., 1989). The cDNA library, like CpTI, was prepared from commercially available cowpeas. A 550-base-pair long AluI–ScaI restriction fragment containing the entire coding sequence for the mature protein, a long leader sequence and the majority of the 3′-non-translated sequence was transferred to the SmaI site of *Agrobacterium*

Table 2. Insecticidal efficacy of cowpea trypsin inhibitors (CpTI) in artificial diets

Insect order	Species	Primary plant hosts
Coleoptera	*Callosobruchus maculatus*	cowpea, and other pulses
	Tribolium confusum	cereal flours
	Anthonomus grandis	cotton, wild species of *Gossypium*
	Diabrotica undecimpunctata	maize, groundnuts, cucurbits
	Costelytra zealandica	clover
Lepidoptera	*Heliothis virescens*	cotton, tobacco, tomato
	Heliothis zea	maize, cotton, tomato (polyphagous)
	Spodoptera littoralis	cotton, cereals (polyphagous)
	Chilo partellus	sorghum, maize, millet, rice, Graminae
Orthoptera	*Locusta migratoria*	polyphagous, preference for Graminae

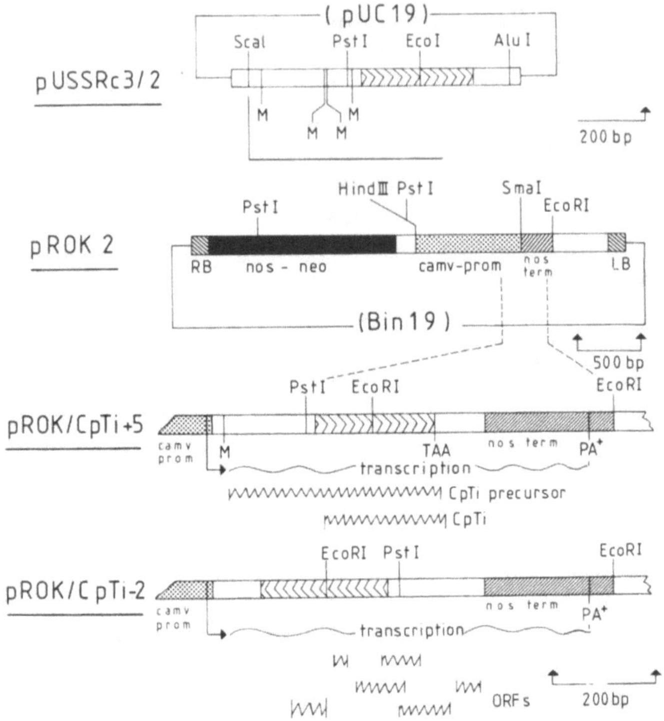

Fig. 1. Construction of a CpTI expression vector for plant transformation (Hilder et al., 1987). The CpTI cDNA pUSSRc3/2, containing a complete mature CpTI coding sequence (> > >; M, in frame initiator codons) was restricted with AluI and ScaI and ligated into the SmaI site of the expression vector pROK 2. Clones with the coding sequence in the correct orientation relative to the promoter (pROK/CpTI + 5) and in the incorrect orientation (pROK/CpTI-2) were generated. Transcripts generated by the clone with the CpTI coding sequence in the correct orientation will be translated to produce a CpTI precursor polypeptide; transcripts from the clone with the CpTI coding sequence in the incorrect orientation contain six short open reading frames

tumefaciens Ti plasmid binary vector, pROK 2 (Baulcombe et al., 1986). This placed the cowpea inhibitor sequence under the control of a strong constitutive promoter derived from cauliflower mosaic virus (Guilley et al., 1982) and the nopaline synthase gene-transcription termination sequence (Bevan et al., 1983). Constructs were identified which contained an insert in the correct orientation relative to the CaMV promoter to produce CpTI (pROK/CpTI + 5) and in the 'reverse' orientation (pROK/CpTI-2), which has six short open reading frames with no identifiable features. This 'reversed' construct was used to produce control transformants. In the present study the crop plant chosen for transformation was tobacco (*Nicotiana tabacum* c.v. Samsun). The constructs were mobilized into

A. tumefaciens (Bevan, 1984) and used to transform leaf discs of *N. tabacum.* The transformants were selected by their antibiotic resistance to kanamycin, and transformed plants were regenerated from shootlets by transfer to a root-inducing, kanamycin-containing agar medium (Horsch et al., 1985). Rooted plants were grown on in soil-based compost.

The presence and levels of CpTI production in the original transformants were measured by dot-immunobinding assays (Jahn et al., 1984) using polyclonal antibodies raised in rabbits against total CpTI. The level of expression in young leaves from different individual pROK CpTI+5 transformants ranged from below the limit of detection to ~1% of total soluble protein. No CpTI expression could be detected in the pROK CpTI-2 transformants, i.e., when the gene was inserted in the incorrect orientation no inhibitor was produced (Hilder et al., 1987). Western blotting of soluble leaf proteins from CpTI expressing transformants showed that polypeptides produced and processed in the transformants corresponded to one of the isoinhibitors present in the cowpea seed; no corresponding polypeptides were produced in the control transformants. Other plant proteins have also been shown to be correctly processed in transgenic tobacco plants (Ellis et al., 1988). The functional integrity of the CpTI produced in these transformed tobacco plants was demonstrated by in vitro trypsin inhibitor activity assay and the lack of complication in obtaining high levels of expression of functional CpTI in tobacco illustrates the advantage of expressing plant proteins in transgenic plants; there are no problems with codon usage, mRNA stability, protein processing, etc. Thus the transformed tobacco plants were able to express the foreign CpTI gene and produce an active trypsin inhibitor whose levels of expression in the leaves of the highest expressing plants were similar to those present in the mature seeds of the resistant variety of cowpea, TVu 2027.

Bioassays were carried out on both sets of transgenic tobacco plants, i.e., on plants containing the gene in the correct orientation and on those containing the gene in the 'reverse' orientation (the latter group of plants formed the control plants) to measure their respective levels of resistance or tolerance. In the first instance the lepidopteran *Heliothis virescens* (tobacco budworm) was tested since it is classified as a serious economic field pest of many crops, including tobacco. Young plants were infested with newly emerged larvae and then sealed into individual plantaria and maintained under controlled environmental regimes. After a trial period of seven days all larvae (both dead and surviving) were removed, their size recorded and the extent of leaf damage measured by computer aided image analysis. The results clearly showed that those CpTI transformants which expressed the foreign protein at approximately 1% were relatively resistant to attack compared to control plants. Some of the transformants showing enhanced levels of insect resistance and some of the control plants were replicated as stem cuttings (Baulcombe et al., 1986) to provide sets of genetically identical plants on which statistically sound insect feeding trials could be run. These further trials provided convincing evidence that the CpTI-producing plants were much more resistant to insect attack. Control plants were devastated by this level of infestation; in trials which we ran beyond seven days, i.e., to 'termination' these control plants were reduced to a stalk. However, on the CpTI-producing plants, although the larvae begin to feed and do some very limited damage to the leaves, they either die or fail to develop as they would on control plants (Fig. 2). In addition to transformation being carried out with *N. tabacum*, CpTI expressing plants of the species *N. plumbaginifolia* were also produced, and as with *N. tabacum*, high CpTI expressors showed very similar results when infested with neonate larvae of the same insect species, i.e., when infested with *H. virescens* (Fig. 3). The observation that larvae did start to feed on the CpTI transformed leaf tissue but readily died or failed to develop is consistent with one of the mechanisms of CpTI toxicity proposed by Gatehouse and Boulter (1983) relying upon a finely controlled balance within the host plant which has to make sufficient nutrients for itself but insufficient to maintain predation, thus the larvae die of starvation at a very early stage. Although the initial bioassays of the transformed plants were carried out using *H. virescens*, trials were subsequently carried out using other lepidopteran pests such as *H. zea* (corn earworm) *Spodoptera littoralis* (armyworm) and *Manduca sexta* (tomato and tobacco hornworm) (Table 3). In all cases the CpTI transformants were resistant to attack compared to control plants. Unfortunately, it was not possible to carry out trials on these plants using any members of the Coleoptera which we are interested in since they would not attack the

a b

Fig. 2. Exposure of control and CpTI-expressing transgenic *Nicotiana tabacum* plants to larvae of *Heliothis virescens* (tobacco, tomato budworm). a, control showing almost complete destruction; b, transgenic CpTI expressor

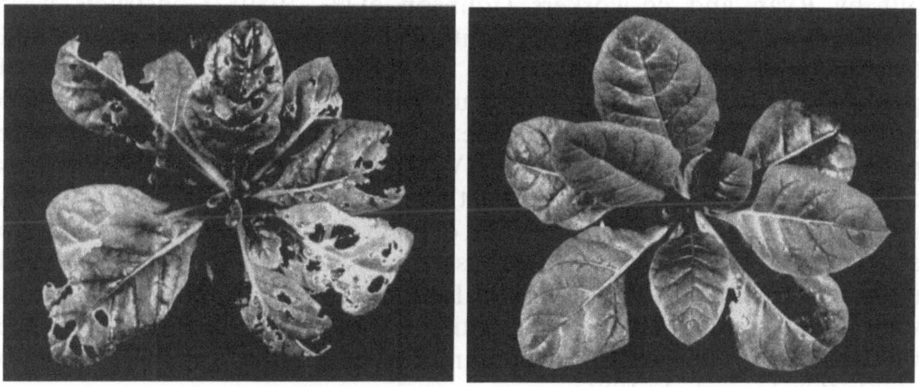

a b

Fig. 3. Exposure of control and CpTI-expressing transgenic *Nicotiana plumbaginifolia* plants to larvae of *Heliothis virescens* (tobacco, tomato budworm). a, control showing significant leaf damage; b, transgenic CpTI expressor showing virtually no leaf damage

Table 3. Insecticidal efficacy of cowpea trypsin inhibitor in transgenic *Nicotiana tabacum*

Insect order	Insect species	Primary plant hosts
Lepidoptera	*Heliothis virescens*	cotton, tobacco, tomato
	Heliothis zea	maize, cotton, tomato (polyphagous)
	Manduca sexta	tobacco, tomato
	Spodoptera littoralis	cotton, cereals (polyphagous)
	Autographa gamma	brassica, potatoes, cereals, legumes (polyphagous)

control plants; in these instances the only information available on the toxicity of CpTI is from artificial diets.

The CpTI gene was shown to be stably inherited through subsequent generations as was the insect resistant phenotype (Hilder et al., 1990). An extensive study under controlled conditions in the plant growth chamber demonstrated that effective levels of CpTI were produced in the transgenic plants without a concomitant 'yield penalty' (Hilder et al., 1989) although as always, caution is required in extrapolating from laboratory to field conditions.

The other successful example where protease inhibitor genes have been used to produce insect resistant transgenic plants has recently been carried out by Ryan and co-workers (Johnson et al., 1989). Constructs were prepared whereby the plasmids contained either the coding region and terminator of tomato Inhibitor I or the coding region and terminator of potato Inhibitor II-k. A third construct carrying a cDNA clone of tomato proteinase Inhibitor II was also made. In all three constructs the genes were placed under the control of the CaMV 35s promoter. As with the plasmids containing the CpTI genes, uptake of these gene constructs into tobacco was mediated via *Agrobacterium* infection of plant tissue (Horsch et al., 1985; An et al., 1986). Leaves from the transgenic tobacco plants were assayed immunologically for the presence of Inhibitors I and II and several plants were found to contain over 200 μg/g tissue; these levels are within the range that is routinely induced by wounding leaves of either tomato or potato plants (Graham et al., 1986).

In order to test the transgenic tobacco plants for insect resistance, first instar larvae of the lepidopteran *Manduca sexta* were placed in petri dishes containing freshly excised tobacco leaves, and each day the larvae were transferred to new dishes containing fresh leaves. The results clearly showed that the presence of the foreign tomato or potato Inhibitor II in tobacco leaves at levels over 100 μg/g tissue severely retarded growth of larvae that fed on them, compared to larvae fed on untransformed plants or transformed plants that did not express these inhibitor genes. At lower levels ($\sim$50 μg/g tissue) larval growth was retarded to a lesser degree than at the higher levels so leading the authors to suggest that there is a dose-dependent relationship between the levels of Inhibitor II and larval growth. In addition to retarding larval growth the presence of Inhibitor II appeared to reduce the amount of leaf tissue consumed.

Interestingly, leaves from transgenic tobacco plants containing up to 130 μg/g of tomato Inhibitor I supported a similar rate of larval growth as those fed control leaves. Since this protein, which is a strong inhibitor of chymotrypsin but only weakly inhibits trypsin, appears to be properly expressed and processed in the transgenic tissue, it would suggest that it is the trypsin inhibitory activity present in Inhibitor II which is primarily responsible for the detrimental effects upon larval growth of *Manduca sexta*. Studies that we have carried out have revealed that the major digestive proteolytic activity in another lepidopteran insect, *Heliothis virescens* was trypsin-like, and that there was virtually no chymotrypsin-like activity present (unpubl. data). Transformation studies using the CpTI gene involved a gene encoding a trypsin/trypsin inhibitor as opposed to the other isoform present in cowpea cotyledons namely, trypsin/chymotrypsin.

From the evidence presented on the use of protease inhibitor genes to date, it would appear that this class of insecticidal protein offers great potential in the production of insect resistant crops. In both examples cited similar strategies have been employed, namely the identification and subsequent exploitation of gene products thought to be involved in natural insect resistance.

VI. References

An G, Watson BD, Chiang CC (1986) Transformation of tobacco, tomato, potato and *Arabidopsis thaliana* using a binary Ti vector system. Plant Physiol 81: 301–305

Applebaum SW, Birk Y, Harpaz I, Bondi A (1964) Comparative studies on proteolytic enzymes of *Tenebrio molitor* L. Comp Biochem Physiol 11: 85–103

Applebaum SW, Konijn AM (1966) The presence of a *Tribolium*-protease inhibitor in wheat. J Insect Physiol 12: 665–669

Barton KA, Whiteley HR, Yang N-S (1987) *Bacillus thuringiensis*-endotoxin expressed in *Nicotiana tabacum* provides resistance to lepidopteran insects. Plant Physiol 85: 1103–1109

Baulcombe DC, Saunders GR, Bevan M, Mayo MA, Harrison BD (1986) Expression of biologically active viral satellite RNA from the nuclear genome of transformed plants. Nature 321: 446–449

Berardi LC, Goldblatt LA (1980) Gossypol. In: Liener IE (ed) Toxic constituents of plant foodstuffs, 2nd edn. Academic Press, New York, pp 183–237

Bevan M (1984) Binary *Agrobacterium* vectors for plant transformation. Nucleic Acids Res 12: 8711–8721

Bevan M, Barnes WM, Chilton MD (1983) Structure and transcription of the nopaline synthase gene region of T-DNA. Nucleic Acids Res 11: 369–385

Birk Y, Gertler A, Khalef S (1963) Separation of a *Tribolium*-protease inhibitor from soybeans on a calcium phosphate column. Biochim Biophys Acta 67: 326–328

Broadway RM, Duffy SS (1986) Plant proteinase inhibiitors: mechanism of action and effect on the growth and digestive physiology of larval *Heliothis zea* and *Spodoptera exigua*. J Insect Physiol 32: 827–833

Campos FAP, Richardson M (1983) The complete amino acid sequence of the bifunctional α-amylase/trypsin inhibitor from seeds of ragi (Indian finger millet, *Eleusine coracana* Gaertn.). FEBS Lett 152: 300–304

Cleveland TE, Thornburg RW, Ryan CA (1987) Molecular characterization of a wound-inducible inhibitor I gene from potato and the processing of its mRNA and protein. Plant Mol Biol 8: 199–207

Dulmage HT (1981) Insecticidal activity of isolates of *Bacillus thuringiensis* and their potential for pest control. In: Burges HD (ed) Microbial control of pests and plant diseases 1970–1980. Academic Press, New York, pp 193–222

Ellis JR, Shirsat AH, Hepher A, Yarwood JN, Gatehouse JA, Croy RRD, Boulter D (1988) Tissue specific expression of a pea legumin gene in seeds of *Nicotiana plumbaginifolia*. Plant Mol Biol 10: 203–214

Feeny PP (1976) Plant apparency and chemical defence. Recent Adv Phytochem 10: 1–40

Fischhoff DA, Bowdish KS, Perlak FJ, Marrone PG, McCormick SM, Niedermeyer JG, Dean DA, Kusano-Kretzmer K, Mayer EJ, Rochester DE, Rogers SG, Fraley RT (1987) Insect tolerant transgenic tomato plants. Biotechnology 5: 807–813

Gatehouse AMR, Boulter D (1983) Assessment of the anti-metabolic effects of trypsin inhibitors from cowpea (*Vigna unguiculata*) and other legumes on development of the bruchid beetle *Callosobruchus maculatus*. J Sci Food Agricult 34: 345–350

Gatehouse AMR, Hilder VA (1988) Introduction of genes conferring insect resistance. In: Proceedings of Brighton Crop Protection Conference, vol 3. Lavenham Press, Suffolk, UK, pp 1234–1254

Gatehouse AMR, Gatehouse JA, Dobie P, Kilminster AM, Boulter D (1979) Biochemical basis of insect resistance in *Vigna unguiculata.* J Sci Food Agricult 30: 948–958

Gatehouse AMR, Gatehouse JA, Boulter D (1980) Isolation and characterisation of trypsin inhibitors from cowpea. Phytochemistry 19: 751–756

Gatehouse JA, Hilder VA, Gatehouse AMR (1990) Genetic engineering of plants for insect resistance. In: Grierson D (ed) Plant genetic engineering. Blackie & Son, Glasgow, pp 105–135 (Plant biotechnology series, vol 1)

Gennis LS, Cantor CR (1976) Double-headed protease inhibitors from black-eyed peas. I. Purification of two new protease inhibitors and the endogenous protease by affinity chromatography. J Biol Chem 251: 734–740

Graham JS, Hall G, Pearce G, Ryan CA (1986) Regulation of synthesis of proteinase inhibitors I and II mRNAs in leaves of wounded tomato plants. Planta 169: 399–405

Green TR, Ryan CA (1972) Wound-induced proteinase inhibitor in plant leaves: a possible defense mechanism against insects. Science 175: 776–777

Guilley H, Dudley RK, Jonard G, Balazs E, Richards KE (1982) Transcription of cauliflower mosaic virus DNA: detection of promoter sequences and characterisation of transcripts. Cell 30: 763–773

Harmsen R, Bliss FA, Cardona C, Posso CE, Osborn TC (1988) Transferring genes for arcelin protein from wild to cultivated beans: implications for bruchid resistance. Annu Rep Bean Improvement Coop 31: 54–55

Hilder VA, Gatehouse AMR, Sheerman SE, Barker F, Boulter D (1987) A novel mechanism of insect resistance engineered into tobacco. Nature 330: 160–163

Hilder VA, Barker RF, Samour RA, Gatehouse AMR, Gatehouse JA, Boulter D (1989) Protein and cDNA sequences of Bowman-Birk protease inhibitors from the cowpea (*Vigna unguiculata* Walp.). Plant Mol Biol 13: 701–710

Hilder VA, Gatehouse AMR, Boulter D (1989) Potential for exploiting plant genes to genetically engineer insect resistance, exemplified by the cowpea trypsin inhibitor gene. Pestic Sci 27: 165–171

Hilder VA, Gatehouse AMR, Boulter D (1990) Genetic engineering of crops for insect resistance using genes of plant origin. In: Lycett GW, Grierson D (eds) Genetic engineering of crop plants. Butterworth, Guildford, pp 51–66

Horsch RB, Fry JE, Hoffman NL, Eichholtz D, Rogers SG, Fraley RT (1985) A simple and general method for transferring genes into plants. Science 227: 1229–1231

Jahn R, Schiebler W, Greengard P (1984) A quantitative dot-immunobinding assay for proteins using nitrocellulose membrane filters. Proc Natl Acad Sci USA 81: 1684–1687

Johnson R, Narvaez J, An G, Ryan CA (1990) Expression of proteinase inhibitors I and II in transgenic tobacco plants: effects on natural defense against *Manduca sexta* larvae. Proc Natl Acad Sci USA 86: 9871–9875

Laskowski M Jr, Sealock RW (1971) Protein proteinase inhibitors—molecular aspects. In: Boyer P (ed) The enzymes, vol 3. Academic Press, New York pp 375–473

Lipke H, Fraenkel GS, Liener IE (1954) Effect of soybean inhibitors on growth of *Tribolium confusum.* J Agricult Food Chem 2: 410–415

Mickel CE, Standish J (1947) Susceptibility of processed soy flour and soy grits in storage to attack by *Tribolium castaneum* (Herbst). Univ Minn Agricult Exp Stat Tech Bull 178: 1–20

Odani S, Ikenaka T (1972) Studies on soybean trypsin inhibitors. IV. Complete amino acid sequence and the anti-proteinase sites of Bowman-Birk soybean proteinase inhibitor. J Biochem 71: 839–848

Plunkett G, Senear DF, Zuroske G, Ryan CA (1982) Proteinase inhibitors I and II from leaves of wounded tomato plants: purification and properties. Arch Biochem Biophys 213: 463–472

Redden RJ, Dobie P, Gatehouse AMR (1983) The inheritance of seed resistance to *Callosobruchus maculatus* F. in cowpea (*Vigna unguiculata* L. Walp.). I. Analysis of parental, F_1, F_2, F_3 and backcross seed generations. Aust J Agricult Res 34: 681–695

Richardson M (1977) The proteinase inhibitors of plants and microorganisms. Phytochemistry 16: 159–169

Ryan CA (1983) Insect induced chemical signals regulating natural plant protection responses. In: Denno RF, McClure MS (eds) Variable plants and herbivores in natural and managed systems. Academic Press, New York, pp 43–60

Sacchi VF, Parenti P, Hanozet GM, Giordanda B, Lutly P, Wolfersberger MG (1986) *Bacillus thuringiensis* toxin inhibits K^+-gradient-dependent amino acid transport across the brush border membrane of *Pieris brassicae* midgut cells. FEBS Lett 204: 213

Seidl DS, Liener IE (1971) Identification of the trypsin-reactive site of the Bowman-Birk soybean inhibitor. Biochim Biophys Acta 251: 83–93

Shivaraj B, Pattabiraman TN (1981) Natural plant enzyme inhibitors. Characterisation of an unusual α-amylase/trypsin inhibitor from ragi (*Eleusine coracana* Geartn.). Biochem J 193: 29–36

Shumway LK, Rancour JM, Ryan CA (1970) Vacuolar protein bodies in tomato leaf cells and their relationship to storage of chymotrypsin inhibitor I protein. Planta 93: 1–14

Shumway LK, Yang VV, Ryan CA (1976) Evidence for the presence of proteinase inhibitor I in vacuolar protein bodies of plant cells. Planta 129: 161–165

Steffens R, Fox FR, Kassel B (1978) Effect of trypsin inhibitors on growth and metamorphosis of corn borer larvae *Ostrinia nubilalis* (Hubner). J Agricult Food Chem 26: 170–174

Vaeck M, Reynaerts A, Hofte H, Jansens S, De Beuckeleer MD, Dean C, Zabeau M, Van Montagu MV, Leemans J (1987) Transgenic plants protected from insect attack. Nature 328: 33–37

Walker-Simmons M, Ryan CA (1977) Immunological identification of proteinase inhibitors I and II in isolated tomato leaf vacuoles. Plant Physiol 60: 61–63

Xavier-Filho J, Campos FAP, Ary MB, Silva CP, Carvalho MMM, Macedo MLR, Lemos FJA, Grant G (1989) Poor correlation between levels of proteinase inhibitors found in seeds of different cultivars of cowpea (*Vigna unguiculata*) and the resistance/susceptibility to predation by *Callosobruchus maculatus*. J Agricult Food Chem 37: 1139–1143

Zwilling R (1968) Evolution of the endopeptidases IV. Alpha- and beta- protease of *Tenebrio molitor* Z Physiol Chem 349: 326–332

Chapter 5

Engineering Microbial Herbicide Detoxification Genes in Higher Plants

Bruce R. Lyon

CSIRO, Division of Plant Industry, GPO Box 1600, Canberra, ACT 2601, Australia

With 5 Figures

Contents

I. Introduction

Although less than ten years have elapsed since the first demonstration of the techniques which enable us to transfer or genetically engineer herbicide resistance into crop plants, there already exists unprecedented interest from molecular biologists, weed scientists, agrochemical manufacturers, plant breeders and farmers alike, in the potential of this new technology. On offer to the molecular biologist is the chance to utilize an assortment of new skills, coupled with the extensive knowledge of herbicide action gathered over the last four decades by weed scientists and biochemists, to address practical problems in agriculture. Agrochemical companies look to this technology to provide the opportunity to increase market share or total sales of herbicides through the wedding of their products to a wider range of crop varieties, while the plant breeders view herbicide resistant traits as a bonus in efforts to differentiate their particular hybrids and varieties from those of their competitors. Finally, there will be obvious advantages for the farmer in using currently available herbicides at greater margins of safety and on additional crops, and in simplifying the integration and rotation of those crops by keeping the hazards of herbicide residue and spray drift toxicity to a minimum.

Significant among the reasons for the investment in genetically engineered herbicide resistant crops is the increasing cost in time and money required to develop new herbicides and have them approved by national regulatory bodies. Thus, the desire to widen the applicability of the older herbicides by creating tolerant crops goes hand in hand with the trend towards fewer and more expensive new generation herbicides. There remains, however, a valid environmental concern with a policy which encourages the continued or expanded use of herbicides which may be questionable with regard to their toxicity and persistence in soil and groundwater. An approach which could conceivably overcome some of these objections involves the co-opting of natural systems of herbicide degradation which have evolved, or are in the process of evolving, due to the sustained use of chemicals in agriculture. Bacterial populations are a prime source of genes which, when transferred into plants, may detoxify herbicides and provide resistance, while these same communities could ensure that the herbicides do not remain in the field for any length of time.

There have been a number of excellent reviews published in recent years on the promise of genetically-engineered herbicide-resistant crops and the progress achieved so far with quite diverse methods of generating tolerance (Comai and Stalker, 1986; Botterman and Leemans, 1988; Shah et al., 1988; Mazur and Falco, 1989; Schulz et al., 1990). In this chapter, I would specifically like to explore the ways in which biotechnology is attempting to harness the processes of microbial herbicide degradation and detoxification for the dual role of producing herbicide tolerant crops and reducing the polluting effects of herbicide usage.

II. Weeds, Agriculture, and Herbicides

Depending on your point of view, a weed can be described as "a plant growing where it is not desired" or "a plant whose virtues have not yet been discovered". The former definition can of course include the residue of a previous crop in a rotation system, but in general, weeds possess a catalogue of traits which enable them to successfully colonize a wide range of environments. These include broad germination requirements, discontinuous germination, seed longevity, rapid growth to flowering, self fertility, unspecialized or wind pollination, continuous seed production, very high seed output, widely dispersed seeds, vegetative growth and competitive factors such as rosette formation, choking growth and the production of allelochemicals (Combellack, 1989; Keeler, 1989). Weeds affect our lives in numerous ways from causing poisonings and allergies to overgrowing our natural forests and fouling our waterways. The primary impact of weeds on mankind, however, occurs in agriculture, where they compete with our crop and pasture plants for applied nutrients and water, harbour pests, and diseases, and decrease harvesting efficiency and the quality of produce.

Farming practices often play a substantial role in determining the spectrum and density of any weed population. Weed infestations may arise from a resident population in place when farming begins or from seeds introduced by irrigation water or intermittent flooding, and are generally aggravated by rigid monoculture cropping regimes and poor farm hygiene. The weed control options available to the grower are manifold, ranging from natural and biological methods to mechanical or chemical approaches. The planting of rotation crops and the use of crop residues has long been recognized as an effective natural means of suppressing weed growth, and studies with biological agents such as rust fungi, insects, nematodes or even grazing mammals indicate that these can also offer significant control. In the cropping situation, the stalwart of weed eradication for millenia has been mechanical cultivation, with the hoe still taking its place beside the high technology plough. The evidence now appears overwhelming, however, that modern methods of mechanical cultivation are creating unacceptible stresses on the soil leading to the problems of compaction, degradation and erosion, and many cropping systems are consequently shifting to minimum tillage techniques involving the use of herbicides.

Herbicides have made such an impact in modern agriculture that it is estimated that their complete withdrawal from use would result in a reduction in food production of up to 35% (Combellack, 1989). The success of herbicides is based solely on their selectivity, the characteristic which enables them to kill a myriad of weedy species but not affect the desired crop plant, even though, as in some cases, the weed and crop may belong to the same genus. While there are a handful of broad-spectrum herbicides which exhibit very little selectivity, the use of such chemicals, although significant, must currently be restricted to noncrop areas, to the preparation of fields prior to the emergence of the crop or to special modes of application. Despite the existence of over one hundred different commercially available herbicides, there remain a host of situations in agriculture where herbicides cannot be efficiently implemented for the control of weeds. Many of these situations arise amongst broadleaf crops such as soybeans and cotton because of the lack of suitable 'knockdown' herbicides which can differentiate between the wanted and unwanted dicotyledonous plants.

In the case of cotton production in Australia for example, the annual cost for herbicides in a crop worth US$ 340–380M is US$ 18M, mostly for

Fig. 1. Weeds and cotton production in Australia. (A) A mature specimen of Noogoora burr (*Xanthium occidentale*) may reach 1 m in diameter. (B) Mechanized cultivation between rows of cotton seedlings. (C) Chipping of weeds among young cotton plants. (D) An experiment in semi-mechanized spot-spraying of weeds in cotton (Photo courtesy of "The Australian Cotton Grower")

non-crop and fallow maintenance, seedbed preparation, pre- and post-emergence residual treatments and knockdown sprays for established grasses. Broadleaf weeds such as Noogoora burr (Fig. 1A), which can cause yield losses of 5–16% at a weed density of as little as one plant per 10 m row of cotton, can initially be controlled by cultivation (Fig. 1B), directed spraying or spot spraying (Fig. 1D), but once the cotton is too high to allow passage of machinery, the only alternative is to remove the weeds by hand (Fig. 1C). It is estimated that an additional US$ 8M is spent every year to chip these recalcitrant weeds.

Further difficulties arise in many cropping situations when plants such as wheat and corn, which can tolerate a range of herbicides directed at broadleaf weeds, are to be grown in adjacent fields or on rotation with broadleaf crops. Cotton and sunflowers, for example, can be severely damaged by the drift of 2,4-D and other phenoxy herbicides which are applied to wheat, while the sensitivity of soybeans to the residual herbicide atrazine used on corn makes it hazardous to rotate these two crops. Closely related crop and weed species also create a dilemma because there is often very little difference between the herbicide sensitivity spectrum of the two plants. Nightshades in tomato and potato, weed beet in sugarbeet, and red rice in rice are just a few examples (Goss and Mazur, 1989).

The solution to these kinds of problems seems to lie in either the development of more selective post-emergence herbicides or in the creation of crops which are resistant to the currently available knockdown or broad-spectrum herbicides. As mentioned in the introduction, it is becoming much more difficult to develop new herbicides which meet the tight industry requirements of potent selective activity, low animal toxicity, short environmental persistence and low production cost. In fact, rates of isolation of new herbicides by the traditional method of laboriously screening thousands of xenobiotic compounds have dropped from 1 in 2,000 to 1 in 20,000 over the last three decades, and it remains to be seen whether the emerging technology of tailor made herbicides, whereby herbicides are engineered to complement the target of choice, will improve this figure to a more workable level in the future (Huppatz, 1990).

III. Transgenic Herbicide Resistant Crops

Even when the full panoply of weed control measures is implemented, the loss in crop yield due to the competitive effects of weeds is in the order of 10%, amounting to US$ 4 billion annually in the U.S.A. alone. This obviously leaves plenty of scope for the introduction of novel weed control strategies, such as transgenic herbicide resistant crops, which could result in more comprehensive weed suppression. As we have seen, the kinds of crops that could benefit from genetically engineered herbicide resistance include cotton, sunflower, soybean, sugarbeet, potato, tomato, and rice, but the list is certainly much longer and would have to include corn, wheat, barley, oilseed rape, and certain vegetables (Goss and Mazur, 1989).

The major impediment to the development of herbicide resistance in the above crops remains the technical ability to transform and regenerate viable transgenic plants carrying the gene(s) for the desired trait. Although most of the dicotyledonous plants can be transformed using *Agrobacterium tumefaciens* (Gasser and Fraley, 1989), in many cases it is not yet possible to regenerate transgenic plants from the agronomically important varieties of these species. Consequently, plants with genetically engineered attributes would need to go through the time consuming process of backcrossing with elite lines. Meanwhile, most monocotyledonous plants cannot be transformed by *Agrobacterium tumefaciens* at all, and other transformation technologies more suited to these plants are presently only in the development phase (Gasser and Fraley, 1989; Potrykus, 1989). Resolution of the problems of gene transfer and regeneration in commercial crop varieties will open the way for the full utilization of the available herbicide resistance genes.

The strategies available to the molecular biologist for the creation of herbicide tolerant crops have the common aim of preventing the binding of the herbicide to sensitive target proteins in the cell. In many cases, the target may be a single enzyme involved in a critical metabolic function such as photosynthesis or amino acid biosynthesis, but in a few instances, the herbicide could affect several different targets, such as the binding receptors for plant hormones. Genetic engineering options can be classified under three subheadings, depending on whether they act to dilute the herbicide through overproduction of the target (overproduction), replace the target with a mutant which is no longer susceptible to the herbicide (replacement), or degrade or otherwise detoxify the herbicide before it reaches the target (detoxification).

The principle of overproduction derives from observations on herbicide resistant mutants selected in cell cultures which overproduce the enzyme targeted by the herbicide. While the overproduction exhibited by these mutants results from gene amplification, attempts to duplicate this in transgenic plants have involved the linking of a single copy of the gene to a high expression promoter such as the 35S promoter of cauliflower mosaic virus (CaMV). This strategy has been successful in two instances; with the overexpression of the 5-enolpyruvylshikimate-3-phosphate synthase (EPSPS) gene from petunia which gives tolerance to glyphosate (Shah et al., 1986), and the alfalfa glutamine synthetase (GS) gene in tobacco which mediates resistance to glufosinate (Eckes and Wengemayer, 1987). The production of these resistant plants was dependent upon the fact that, in each case, only a single enzyme was affected by the herbicide and the gene for this enzyme could be readily isolated and characterized.

The technique of replacing the gene for a herbicide sensitive target with a mutant gene that encodes a resistant target has been used to generate herbicide tolerant plants in a number of instances. Tobacco plants showing increased tolerance to glyphosate were produced by transformation with a mutant *Salmonella typhimurium* EPSPS gene linked to a plant promoter which expressed EPSPS in the plant cytoplasm (Comai et al., 1985). Glyphosate resistant petunia plants have also been generated by transformation with a mutant *Escherichia coli* EPSPS gene provided with a chloroplast transit sequence which enables the new enzyme to pass into the chloroplast where EPSPS is predominantly localized (della-Cioppa et al., 1987). A similar approach was adopted to produce atrazine tolerant plants using a mutant gene for the photosynthetic membrane protein QB obtained from *Amaranthus hybridus* (Cheung et al., 1988). In the generation of transgenic plants resistant to the sulfonylurea herbicides, a cloned yeast gene for acetolactate synthase (ALS), the enzyme affected by these herbicides, was used to isolate homologous genes from resistant tobacco and arabidopsis mutants, thereby obviating the need for additional chloroplast transit sequences (Mazur et al., 1987; Haughn et al., 1988). Glufosinate resistant plants have also been created by transformation with a resistant GS gene obtained from a mutant alfalfa, however, differential expression of this gene in the roots and not the leaves of the plant suggests that this gene may encode a tissue-specific isozyme (Mazur and Falco, 1989). The use of replacement as a mechanism for producing herbicide tolerant plants has encountered some problems, not the least of which is the targeting of expression to particular tissues or cellular compartments, and once again, the requirement for considerable knowledge of a (single) target is necessary for a successful result.

In many instances, the selectivity of a herbicide is determined by the ability of the target plant to detoxify the chemical using endogenous enzymes. Resistant plants, unlike their sensitive cousins, possess metabolic pathways containing enzymes such as mixed function oxidases, N-glucosyl transferases and glutathione S-transferases which act to convert the herbicide to an inactive derivative (Comai and Stalker, 1986; Schulz et al., 1990). Such enzyme systems provide a wealth of opportunity for the genetic engineering of herbicide resistant crops because variations in the herbicide-modifying capacity of a plant result in a much more subtle shift in specificity than does, for example, the conversion of the target to a resistant form. Thus, the former plant may only be resistant to herbicides with molecular structures containing a site which can be modified, whereas the latter crop would be resistant to an entire class of herbicides due to their function and not their structure.

Despite the availability of such enzymes in plants, most efforts to genetically engineer detoxification mechanisms into crops have involved the use of degradative enzymes derived from microorganisms. The reasons for this include a broader understanding of microbial degradative pathways, the greater accessibility of microbial genes, and the fact that most interest to date has focused on the non-selective herbicides for which few resistance mechanisms are known in plants. Detoxification strategies based on microbial genes have been successfully used to generate transgenic plants resistant to bromoxynil (Stalker et al., 1988), glufosinate (De Block et al., 1987) and 2,4-D (Streber and Willmitzer, 1989; Lyon et al., 1989), and potential degradative pathways exist for many other important herbicides. The main advantage of this strategy is that it is unnecessary to first identify the site of action of the herbicide in the plant as the aim is to intercept the chemical before it reaches the target. It follows from this that the target also need not be a single protein, as all components of the cell should be protected by the resistance mechanism.

IV. Microbial Degradation of Herbicides

A perusal of Table 1 reveals that microorganisms of one kind or another are implicated in the degradation of an impressive array of herbicides comprising representatives of almost every agrochemical family. The degradation mechanisms are as diverse as the species involved and are all aimed at deriving some benefit for the organism, whether it be the supply of carbon, nitrogen, phosphate or sulphate, from the breakdown of the compound. In some cases, degradation does not lead to the complete mineralization of the compound, with the result that certain by-products can be left as recalcitrant environmental pollutants. Even under these circumstances, however, the herbicide may have undergone chemical alterations that make it either less phytotoxic or more suitable for modification or conjugation by plant detoxification systems. The following sections briefly review some of the mechanisms of microbial breakdown for the major herbicide families.

A. Acetamides

The acetamide or acylanilide herbicides such as alachlor (Lasso), metolachlor (Dual), propachlor (Ramrod) and propanil are generally soil incorporated pre-crop emergence for the control of annual grasses and some broad-leaved weeds in a range of broad-leaved crops and grains. The incomplete mineralization of alachlor by the soil fungus, *Chaetomium globosum*, is believed to occur through the hydrolysis of the amide linkage as illustrated in Fig. 2 (Tiedje and Hagedorn, 1975). The degradation of propachlor by the soil fungus *Fusarium oxysporum*, on the other hand, seems to procede via dehalogenation, and produces compounds which are much less toxic to oat seedlings (Kaufman and Blake, 1973). Metolachlor is more persistent than the other acetamides, but has been shown to undergo dealkylation similar to alachlor (McGahen and Tiedje, 1978) or hydroxylation and demethylation (Krause et al., 1985; Saxena et al., 1987). Propanil is rapidly inactivated on contact with soil and can be completely metabolized by a *Pseudomonas* sp. capable of degrading the aromatic ring (Zeyer and Kearney, 1982).

B. Carbamates

The carbamates have a similar spectrum of activity to the acetamides and their structures, although less complex, are analogous. Chlorpropham (CIPC) and propham (IPC) are degraded by a variety of organisms including the blue-green bacterium *Anacystis nidulans* (Wright and Maule, 1982) and several *Pseudomonas* spp. (Kaufman and Blake, 1973; Vega et al., 1985; Marty and Vouges, 1987). In *Pseudomonas alcaligenes*, hydrolysis of chlorpropham to the less phytotoxic metabolite 3-chloroaniline (Fig. 2) is achieved by an amidase with a molecular weight of 68,000 (Marty and Vouges, 1987).

C. Dinitroanilines

The dinitroanilines are usually soil incorporated for residual control of annual and perennial broad-leaved weeds and grasses in cotton, soybeans, cereals and legumes. Under aerobic conditions, trifluralin (Treflan) is initially degraded to less phytotoxic compounds by dealkylation followed by progressive reduction of the nitro groups to create inactive metabolites (Probst and Tepe, 1969) (Fig. 2). There is little evidence to suggest that trifluralin can be completely mineralized by soil microorganisms.

D. Diphenyl Ethers

The diphenyl ethers such as oxyfluorfen (Goal) are used for the pre- or post-emergence control of annual broad-leaved weeds and grasses in cereals and in cotton and soybeans, where they can cause injury if incorrectly applied. These compounds are extremely recalcitrant to biodegradation, but a recent paper reports the isolation of a diphenyl ether cleavage gene from *Erwinia* sp. which may make these compounds more accessible to further breakdown (Liaw and Srinivasan, 1989) (Fig. 2). This gene has been cloned and expressed in *Escherichia coli* where it makes a polypeptide of molecular weight 21,000.

Table 1. Environmental persistence and microbial degradation of herbicides[a]

Chemical family	Herbicide	Persistence (half-life*)	Degradative microorganisms	References
Acetamide	Alachlor	6–10 wk	Sewage + freshwater microbes, anaerobic stream sediment	Novick and Alexander, 1985; Bollag et al., 1986
	Metolachlor	*15–50 d	Actinomycete, bacteria + fungi	Krause et al., 1985; Saxena et al., 1987
	Propachlor	4–6 wk	Sewage + freshwater microbes, soil microbes	Novick and Alexander, 1985; Novick et al., 1986
	Propanil	1–3 d	*Pseudomonas* sp., blue-green bacteria + algae	Zeyer and Kearney, 1982; Wright and Maule, 1982
Carbamate	Chlorpropham	4–11 wk	Blue-green bacteria; *Pseudomonas cepacia*	Wright and Maule, 1982; Vega et al., 1985
	Desmedipham	*30–60 d	Bacteria, fungi + yeast	Knowles and Benezet, 1981
	Phenmedipham	*25 d	Bacteria, fungi + yeast	Knowles and Benezet, 1981
	Propham	*5–15 d	Sewage microbes	Wang et al., 1984
Dinitroaniline	Profluralin		Bacteria	Stralka and Camper, 1981
	Trifluralin	26–35 wk	Soil microbes	Zeyer and Kearney, 1983
Diphenyl ether	Oxyfluorfen	4–8 wk	*Erwinia* sp.	Liaw and Srinivasan, 1989
Nitrile	Bromoxynil	*10 d	*Klebsiella ozaenae*	McBride et al., 1986
Organophosphorus	Glufosinate	*30–40 d	*Rhodococcus* sp.	Bartsch and Tebbe, 1989
	Glyphosate	8 wk	*Pseudomonas* sp.; *Flavobacterium* sp.; *Arthrobacter* sp.	Moore et al., 1983; Balthazor and Hallas, 1986; Pipke et al., 1987

Phenoxy	2,4-D	6 wk	*Alcaligenes eutrophus*	Don and Pemberton, 1981
			Alcaligenes sp.	Amy et al., 1985
	2,4,5-T		*Pseudomonas cepacia*	Kilbane et al., 1982
	MCPA	13–17 wk	Soil microbes	Lappin et al., 1985, Smith and Hayden, 1981
Pyridilium	Paraquat		Yeast	Carr et al., 1985
Sulfonylurea	Chlorsulfuron	4–6 wk	*Streptomyces griseolus*	O'Keefe et al., 1988
	Sulfometuron-methyl	*28 d	*Streptomyces griseolus*	O'Keefe et al., 1988
Thiocarbamate	Diallate	*30 d	Soil microbes	Anderson, 1984
	Triallate		Soil microbes	Anderson, 1984
	EPTC		Bacteria + fungi	Lee, 1984
			Arthrobacter sp.	Tam et al., 1987
Triazine	Atrazine	22–30 wk	Fungi	Giardina et al., 1982
			Pseudomonas sp.	Behki and Khan, 1986
	Ametryn		Bacteria	Cook and Hutter, 1982
	Prometryn	4–12 wk	Bacteria	Cook and Hutter, 1982
Uracil	Bromacil	30 wk	*Pseudomonas* sp.	Chaudhry and Cortez, 1988
Urea	Diuron	17–34 wk	Sediment microbes	Attaway et al., 1982

[a] See MacRae (1989), for a more comprehensive list of degradative microorganisms

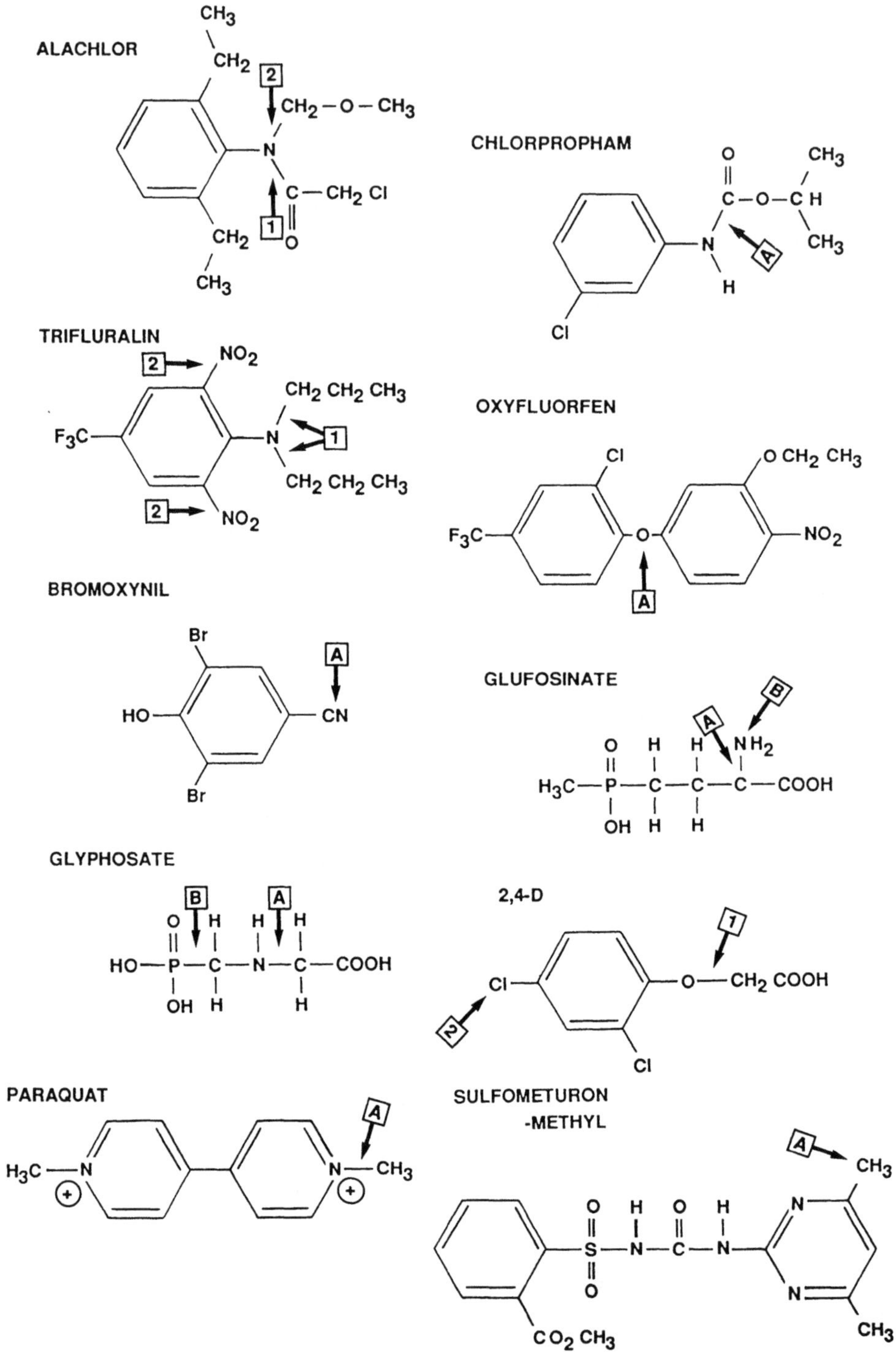

ALACHLOR
CHLORPROPHAM
TRIFLURALIN
OXYFLUORFEN
BROMOXYNIL
GLUFOSINATE
GLYPHOSATE
2,4-D
PARAQUAT
SULFOMETURON
-METHYL

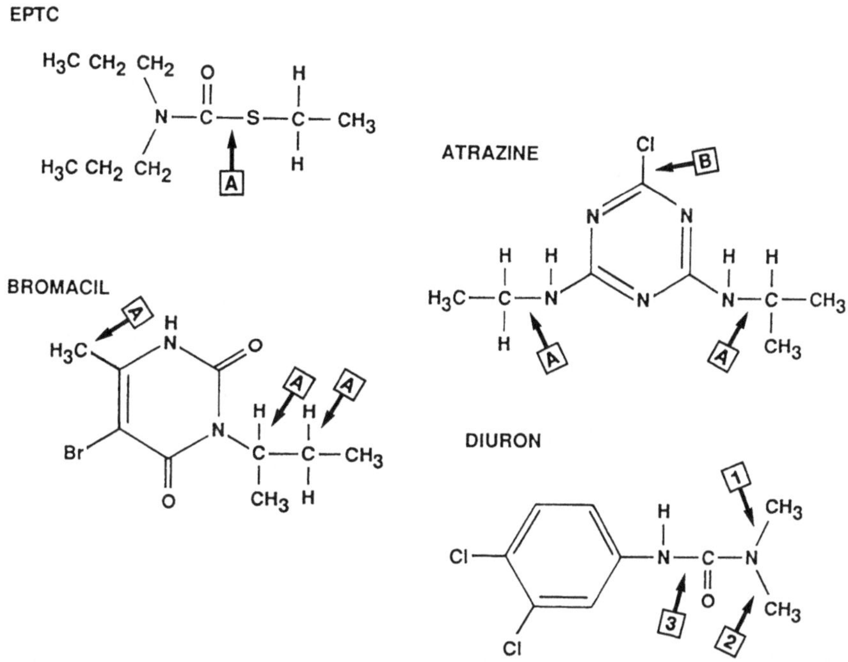

Fig. 2. Representative members of major herbicide families and the enzymes involved in the initial stages of microbial degradation. Arrowed letters refer to alternative mechanisms of breakdown, arrowed numerals refer to sequential steps in breakdown. Alachlor: 1, amidase; 2, hydrolase. Chloropropham: A, amidase. Trifluralin: 1, aminase/dealkylase; 2, reductase. Oxyfluorfen; A, hydrolase. Bromoxynil: A, nitrilase. Glufosinate: A, oxidase; B, acetylase. Glyphosate: A, hydrolase; B, carbon-phosphorus lyase. 2,4-D: 1, monooxygenase; 2, hydroxylase. Paraquat: A, demethylase. Sulfometuron-methyl: A, hydroxylase. EPTC: A, esterase. Atrazine: A, aminase/dealkylase; B, hydroxylase. Bromacil: A, hydroxylase. Diuron: 1 and 2, demethylase; 3, amidase

E. Nitriles

This family includes bromoxynil (Buctril) which is employed for the post emergent suppression of broad-leaved weeds in cereals and grasses. Bromoxynil is degraded to the non-phytotoxic aromatic acid by a nitrilase enzyme (Fig. 2), before being completely mineralized by co-metabolic soil organisms (McBride et al., 1986). The plasmid-encoded gene for the bromoxynil-specific nitrilase of *Klebsiella ozaenae* has been cloned in *Escherichia coli* and sequenced, and the enzyme was found to comprise two identical subunits of molecular weight 38,100 (Stalker and McBride, 1987; Stalker et al., 1988).

F. Organophosphorus Compounds

The organophosphorus herbicides are broad-spectrum herbicides applied to non-crop areas, pre-crop emergence, or by novel methods such as wick-wipers to avoid contact with the crop. Also known as phosphinothricin, the herbicide glufosinate (Basta) is non-toxic to the majority of microorganisms and complete mineralization can occur in non-sterile soils (Smith, 1989). A *Rhodococcus* sp. was found to utilize glufosinate as a sole source of nitrogen by oxidation to the non-phytotoxic oxoacid (Fig. 2), and this organism and several others could detoxify the herbicide by acetylation (Bartsch and Tebbe, 1989) (Fig. 2). Acetylation is also the mechanism used to protect *Streptomyces hygroscopicus*, the organism which produces bialaphos, the precursor of phosphinothricin, and the acetylase gene (*bar*) has been cloned, sequenced and shown to encode a protein of molecular weight 22,000 (Thompson et al., 1987). Glyphosate (Roundup) can be degraded by microorganisms using one of two different pathways which destroy the phytotoxicity of the compound. The more common hydrolysis to produce aminomethylphosphonate (AMPA) (Fig. 2) is carried out by *Flavobacterium* sp. GDI (Balthazor and Hallas, 1986), *Pseudomonas* sp. LBr (Jacob et al., 1988), several other *Pseudomonas* sp. and *Alcaligenes* sp. (Talbot et al., 1984) and *Arthrobacter atrocyaneus* (Pipke and Amrhein, 1988a). Hydrolysis of the carbon-phosphorus (C-P) bond by a C-P lyase to give inorganic phosphate and sarcosine is the method employed by *Pseudomonas* sp. PG2982 (Shinabarger and Braymer, 1986) and *Arthrobacter* sp. GLP-1 (Pipke and Amrhein, 1988b) (Fig. 2). A strain of *Enterobacter aerogenes* has also been isolated which possesses a C-P lyase with limited activity towards glyphosate (Murata et al., 1989), and this enzyme was shown to comprise one subunit of molecular weight 560,000 and two identical subunits of molecular weight 55,000. Both of the above routes are self-limiting due to the inhibitory effect of inorganic phosphate on glyphosate uptake (Pipke et al., 1987). Only the *Flavobacterium* sp. seems capable of degrading glyphosate in the presence of orthophosphate, which is critical if the organism is to be of importance in the clean-up of natural environments (Balthazor and Hallas, 1986).

G. Phenoxys

The phenoxyacetates such as 2,4-dichlorophenoxyacetic acid (2,4-D), 2,4,5-trichlorophenoxyacetic acid (2,4,5-T) and 2-methyl-4-chlorophenoxyacetic acid (MCPA) are used for control of broad-leaved weeds in cereals and non-crop areas, although 2,4,5-T has now been withdrawn from use in many countries. A wide range of bacterial genera including *Acinetobacter*, *Alcaligenes*, *Arthrobacter*, *Corynebacterium*, *Flavobacterium* and *Pseudomonas* have been shown to completely metabolize 2,4-D (cited, in Chaudhry and Huang, 1988), and the best characterized of these is a strain of *Alcaligenes eutrophus* which carries a plasmid encoding the first six enzymes of the degradative pathway (Don et al., 1985). Degradation proceeds through the action of a 2,4-D monooxygenase which generates the less phytotoxic metabolite 2,4-dichlorophenol, followed by a hydroxylation step that yields an inactive product (Fig. 2). The gene for the monooxygenase (*tfdA*) has been shown to encode a protein of molecular weight 32,000 (Streber et al., 1987), while the 2,4-dichlorophenol hydroxylase with a total molecular weight of 224,000 is believed to comprise two non-identical subunits of molecular weight 67,000 and 45,000 (Liu and Chapman, 1984). MCPA is also degraded by this series of enzymes, and microbial communities that can degrade MCPA, MCPB, mecoprop and dichlorprop have been reported (Kilpi, 1980; Lappin et al., 1985; Smith and Hayden, 1981).

H. Pyridiliums

Paraquat (Gramoxone) is a contact herbicide employed for broad-spectrum control of dicotyledonous weeds in orchards and plantations and is often used for the pre-harvest dessication of crops. A yeast (*Lipomyces starkeyi*) is able to degrade paraquat to acetic acid and CO_2 following an initial step that removes the methyl or methylamine group (Carr et al., 1985) (Fig. 2). Paraquat is readily adsorbed to soil, however, and although biologically inert, it can remain unavailable for degradative attack for many years.

I. Sulfonylureas

Some of the sulfonylureas such as chlorsulfuron (Glean) and metsulfuron-methyl (Ally) are selective for use in cereals for the post-emergence control of annual and perennial broad-leaved weeds, while others like sulfometuron-methyl (Oust) are broad-spectrum and are only recommended for non-cropland use. The hydrolysis of sulfometuron-methyl in acid soils followed by the complete degradation of the fragments by microbial action has been reported (Anderson and Dulka, 1985). In *Streptomyces griseolus*, an inducible cytochrome P-450 is implicated as the terminal oxidase in a monooxygenase system responsible in part for the metabolism of sulfonylureas by this organism (O'Keefe et al., 1988). Chlorsulfuron and sulfometuron-methyl are thought to undergo an initial hydroxylation reaction (Fig. 2) which is analogous to the mechanism of detoxification observed in plants (Romesser and O'Keefe, 1986).

J. Thiocarbamates

The thiocarbamate EPTC is soil incorporated in a wide range of broad-leaved crops for the suppression of annual and perennial grasses and some broad-leaved weeds. Soil microorganisms contribute significantly to the degradation of the thiocarbamates, with hydrolysis at the ester linkage being the most likely form of attack (Fang, 1969) (Fig. 2). The isolation of an *Arthrobacter* sp. carrying a plasmid associated with the degradation of EPTC should hasten the elucidation of this pathway (Tam et al., 1987).

K. Triazines

The triazines, including atrazine, prometryn (Gesagard) and simazine (Gesatop), are residual herbicides applied pre- or post-emergence for the control of annual grass and broad-leaved weeds. Soil microorganisms are implicated in the formation of less phytotoxic compounds by dealkylation, and non-phytotoxic compounds by hydroxylation (Knuesli et al., 1969) (Fig. 2). A *Nocardia* sp. was isolated which can use atrazine as a sole carbon and nitrogen source (Giardina et al., 1982), while the complete degradation of this herbicide initiated by a *Pseudomonas* sp. and completed by a *Rhodococcus* sp. has been hypothesized (Behki and Khan, 1986).

L. Uracils

The herbicide bromacil (Hyvar X) is employed for the total control of weeds and brush on non-cropland and selective control in orchards. Bromacil is extremely persistent in soil, with degradation finally proceeding via hydroxylation of the side chain alkyl groups followed by ring opening and complete metabolism (Gardiner et al., 1969) (Fig. 2). A strain of *Pseudomonas* was shown to possess two plasmids associated with the ability to metabolize bromacil as a sole source of carbon and energy (Chaudhry and Cortez, 1988).

M. Ureas

The urea herbicides such as monuron, diuron, and fluometuron (Cotoran) are used for total weed control in non-cropland, or for residual suppression of most weeds from seed in a variety of crops. Degradation proceeds via demethylation to a less phytotoxic metabolite followed by further demethylation and hydrolysis to inactive products (Geissbuhler, 1969) (Fig. 2). Organisms involved in the transformations include *Pseudomonas* sp.,

Xanthomonas sp., *Penicillium* sp., and *Aspergillus* sp., and it is presumed that, with time, complete mineralization results.

One can see from the above that aerobic bacteria with simple nutritional requirements such as *Pseudomonas, Alcaligenes, Klebsiella,* and *Arthrobacter* are frequently implicated in the degradation of herbicides. On the one hand this reflects the extraordinary range of catabolic pathways possessed by these organisms; *Pseudomonas cepacia* alone can utilize over one hundred different compounds as carbon and energy sources (Stanier et al., 1966). Enrichment procedures have, however, been traditionally biased towards the isolation of these genera, and it is suggested that large reserves of degradative capacity may well lie untapped amongst less readily purified bacteria and fungi (Cook et al., 1983).

Degradative microorganisms can be isolated from sewage works and production plant outflows, but agricultural fields that have undergone repeated exposure to herbicides remain the best source of inocula (Cook and Hutter, 1982). Enrichment of the desired organism is routinely achieved by continuous culture in the presence of increasing concentrations of the herbicide, while a basal level of a more readily metabolized carbon source ensures a steadily growing population from which mutants can be selected (Cook et al., 1983). In the isolation of organisms capable of degrading the phenoxy herbicides 2,4-D, MCPA and dichlorprop (Fig. 3) for example, MCPA and vanillate or dichlorprop and benzoate were added as carbon sources with an inoculum of soil from fields treated with mecoprop (Kilpi, 1980). Vanillate was used due to a similarity in structure to MCPA, and the success of this strategy indicates that enzymes involved in the degradation of MCPA may be related to those responsible for the metabolism of vanillate, a product of lignin degradation. Thus, although most herbicides are man-made and lack natural analogues, it is probable that enzymes for herbicide degradation are derived from existing pathways for the metabolism of naturally occurring organic compounds and polymers.

The rapid evolution of new catabolic activities is believed to result from a variety of genetic mechanisms including gene duplication and subsequent mutational divergence, gene recruitment from diverse sources, legitimate or illegitimate genetic recombination, and alterations in regulatory control (Ghosal et al., 1985). In an experiment to test this theory, microorganisms from various waste sites together with characterized strains carrying non-herbicide degradative plasmids were incubated in the presence of high concentrations of the non-herbicide substrates and small quantities of the phenoxy herbicide 2,4,5-T (Kellogg et al., 1981). A strain of *Pseudomonas cepacia* capable of degrading the previously recalcitrant 2,4,5-T evolved presumably by one or more of the above mechanisms following continuous culture for 8–10 months (Kilbane et al., 1982). This procedure has been termed "plasmid-assisted molecular breeding" and offers great potential for the generation of microorganisms with new metabolic functions.

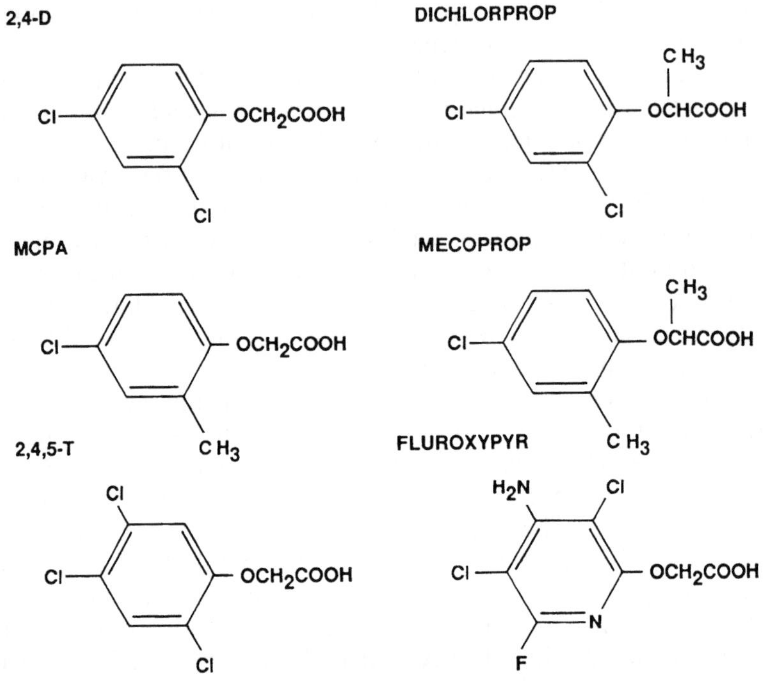

Fig. 3. Six members of the phenoxy family of herbicides. These compounds mimic the action of the plant hormone auxin and are phytotoxic to most broad-leaved plants

Although 2,4,5-T is closely related to other phenoxy herbicides, it is interesting that this compound is not normally degraded by the same catabolic pathways as 2,4-D and MCPA (Fig. 3). It is assumed that the initial enzyme in 2,4-D degradation, the 2,4-D monooxygenase, is highly substrate specific, and does not exhibit activity on 2,4,5-T due to the presence of an extra ring chlorine, nor on the propionic analogues of 2,4-D (dichlorprop) and MCPA (mecoprop) because of the additional methyl group present on the side-chain (Fig. 3). The existence of such tight specificity raises the option of mutating the relevant gene to obtain less rigid substrate requirements so as to deal with compounds such as 2,4,5-T or even fluroxypyr (Fig. 3). As alluded to above, changes in enzyme specificity can evolve following the duplication of catabolic genes, and recent evidence for the presence of a second non-identical copy of the 2,4-D monooxygenase gene on the native degradative plasmid suggests that this process may indeed be occurring in vivo (Perkins and Lurquin, 1988).

Methods associated with protein engineering, such as site-directed mutagenesis and in vitro recombination, offer a more controlled system for the manipulation of genes encoding catabolic enzymes. Much has been learnt in recent years about the catalytic sites of certain enzymes, and it is not inconceivable that highly effective degradative enzymes could one day be designed de novo. In theory then, it should be possible to enhance the natural catabolic ability of microorganisms to efficiently degrade current as well as future herbicides, and eliminate many of the problems associated with herbicide residual toxicity and pollution.

V. Genetic Engineering of Plants

Resistance to three different herbicides has already been engineered in plants through the introduction of genes encoding bacterial detoxification enzymes.

A. Resistance to Bromoxynil

Bromoxynil (Buctril) is a potent inhibitor of the chloroplast photosystem II with good activity on dicotyledonous plants which, unlike many monocots, lack an effective detoxification pathway. Transgenic tobacco plants resistant to high doses of a commercial formulation of bromoxynil were generated by transformation with a gene from *Klebsiella ozaenae* (*bxn*), which encodes a specific nitrilase that converts the cyano moiety of bromoxynil to the non-phytotoxic acid derivative (Stalker et al., 1988) (Fig. 2). The bacterial gene was linked to a promoter derived from a tobacco light-inducible, tissue-specific ribulose biphosphate carboxylase small subunit gene in order to limit expression to the photosynthetic tissue. It was reasoned correctly, that detoxification of bromoxynil in the green tissue should be sufficient for resistance if the primary target was photosynthesis and the herbicide was not translocated. Transgenic *Nicotiana tabacum* cv. Xanthi plants expressing the highest levels of nitrilase (0.01% of total leaf protein) were unaffected by spraying with concentrations of bromoxynil eightfold higher than the recommended field dose. Sprayed plants grew to maturity and set seed normally, and resistance was inherited as a dominant trait in the succeeding generation.

B. Resistance to Phosphinothricin (Glufosinate)

Phosphinothricin is a potent inhibitor of glutamine synthetase, and plants treated with this non-selective herbicide quickly succumb to the toxic effects of accumulated ammonia (Tachibana et al., 1986). Transgenic tobacco, potato and tomato plants that are completely insensitive to phosphinothricin have been generated by transformation with the *bar* gene from *Streptomyces hygroscopicus* (De Block et al., 1987). This gene encodes a phosphinothricin acetyltransferase (PAT) which detoxifies the herbicide by acetylation at the free NH_2 group (Thompson et al., 1987) (Fig. 2).

In its native form, the *bar* gene possesses a GTG translation initiation codon, and in order to guarantee proper translation initiation in plants, this was changed to an ATG by oligomutagenesis. The modified gene was then inserted between the CaMV 35S promoter and the termination and polyadenylation signal of the octopine T-DNA gene 7, cloned into the plant transformation vector pGV1500, and mobilized into *Agrobacterium tumefaciens* strain C58C1(pGV2260). Leaf discs of *Nicotiana tabacum* cv. Petit Havana SR1 were infected with this organism and transformed shoots were selected either on kanamycin or various levels of phosphinothricin. The transgenic tobacco plants were transferred to soil and their growth, together with that of transformed potato (*Solanum tuberosum* cv. Berolina, cv. Bintje, and cv. Desiree) and tomato (*Lycopersicum esculentum* cv. Lukullus) plants produced in a similar manner, was indistinguishable from untransformed control plants.

Glasshouse and/or field spraying tests have revealed that the transgenic plants are resistant to high levels of the commercial preparations of phosphinothricin (Basta) and its plant-activated precursor bialaphos (Herbiace) (De Block et al., 1987; De Greef et al., 1989). In the field trials, transgenic tobacco plants sprayed with between one and four times the regular herbicide dose showed only some slight discolouration 48 h after application (attributed to surfactants in the preparation which are not adapted for selective use on crops) and then recovered completely, whereas untransformed plants were killed by the lowest dose. Interestingly, the expression of PAT in the transgenic tobacco could vary between 0.001% and 0.1% of total extracted protein without adversely affecting the level of herbicide resistance. Transgenic potato plants were not visibly affected by herbicide treatments that killed the controls, although one cultivar suffered a 20% yield loss (measured by tuber weight) when sprayed with four times the normal dose of herbicide. When sprayed with Basta at the normal application rate, the potato cultivars yielded 11–51% more than untreated controls presumably as a result of the more effective weed control.

There has been a recent report that transgenic tobacco and alfalfa plants transformed with a PAT gene from *Streptomyces viridochromogenes* also exhibit protective levels of resistance to phosphinothricin (Donn et al., 1990). In this case the gene was synthesized with a codon usage optimized for expression in plants, and resistance to twice the normal field dose of herbicide was achieved, despite the fact that the alfalfa plants only expressed the PAT gene at lower levels.

C. Resistance to 2,4-D

Two research groups have succeeded in producing transgenic tobacco plants with increased resistance to 2,4-dichlorophenoxyacetic acid (2,4-D) resulting from the expression of a bacterial detoxification enzyme (Streber and Willmitzer, 1989; Lyon et al., 1989). In both cases, the enzyme employed was a 2,4-D monooxygenase which catalyzes the cleavage of the acetate side chain of 2,4-D to give glyoxylate and 2,4-dichlorophenol (Don et al., 1985) (Fig. 2). The latter compound is between 50 and 100 times less toxic to tobacco plants than 2,4-D as measured by leaf disc and seed germination assays. The gene encoding the 2,4-D monooxygenase (*tfdA*) resides on the *Alcaligenes eutrophus* degradative plasmid JP4, and subcloning and sequence analysis has assigned this gene to an 861 bp open reading frame located on a 2.06 kb *Bam*HI–*Sal*I restriction fragment (Streber et al., 1987) (Fig. 4A). The following description of the development and analysis of 2,4-D resistant plants will concentrate on the research conducted in this laboratory.

The native form of the *tfdA* gene bears a GTG translation initiation codon so oligomutagenesis was undertaken to change this codon to an ATG and alter two of the bases immediately upstream to match a published plant consensus sequence for this region (Fig. 4A). The *tfdA* gene was inserted between the CaMV 35S promoter and the nopaline synthase 3′ polyadenylation signal of a plant expression vector and cloned into the binary plasmid pGA470 (Fig. 4B). The binary plasmid construct was then transferred into *Agrobacterium tumefaciens* strain LBA 4404 and this organism was used to transform leaves of *Nicotiana tabacum* cv. Wisconsin 38. Selection for transformed shoots was made on media containing kanamycin as direct selection for resistance to 2,4-D was not successful. Leaf discs taken from regenerated plantlets and plated on media containing 2,4-D were shown to produce shoots on much higher concentrations of the herbicide than leaf tissue from untransformed plants. The transgenic plants were subsequently transferred to soil where they grew to maturity and set seed normally.

F_1 transgenic and wildtype plants were sprayed with several different 2,4-D preparations over a range of concentrations and it was found that plants transformed with the 2,4-D monooxygenase gene were at least 10-fold more resistant to the herbicide than non-transformed plants (Fig. 5A). A field release of these plants is yet to be undertaken, but a spraying trial of field grown wildtype tobacco conducted in tandem with glasshouse spraying of transgenic plants confirmed the increased herbicide tolerance of the latter. Transgenic plants were unaffected by a commercial preparation of 2,4-D when sprayed at a rate which severely damaged control plants and suppressed broadleaf weeds in the field. Transgenic plants showed signs of damage when sprayed at higher levels, however, indicating that there is a need to develop more effective expression of the resistance to provide a margin of crop safety.

Transgenic tobacco seed can germinate on media containing 2,4-D at levels that inhibit the germination of wildtype seed (Fig. 5B). A direct relationship exists between the level of resistance exhibited by the seed and

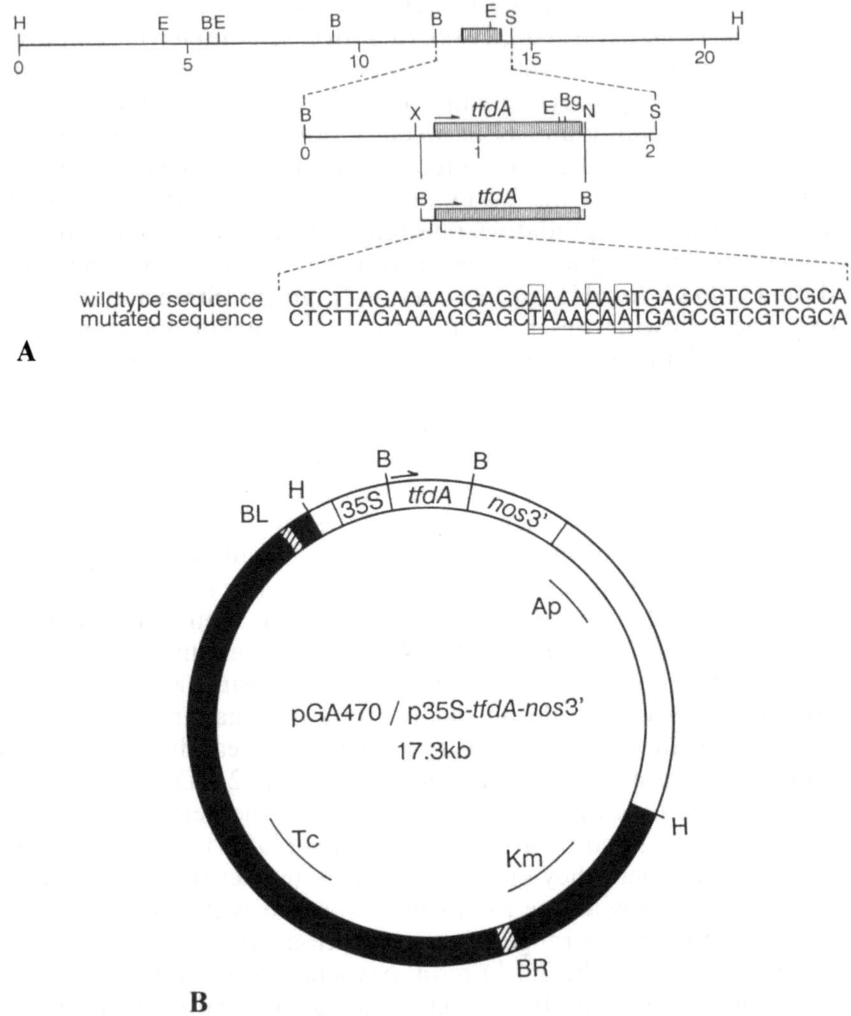

Fig. 4. Cloning of the gene for 2,4-D monooxygenase (*tfdA*) from *Alcaligenes eutrophus* plasmid JP4. (A) the 960 bp *Bam*HI–*Bam*HI fragment was subjected to oligomutagenesis in the vicinity of the bacterial GTG translational start codon (base changes are boxed) to produce a plant consensus sequence and ATG translational start codon (underlined). The scale beneath each map is in kilobase pairs. Restriction endonuclease sites are B, *Bam*HI; Bg, *Bgl*II; E, *Eco*RI; H, *Hin*dIII; N, *Nco*I; S, *Sal*I; X, *Xba*I. The arrow indicates the direction of translation of *tfdA*. (B) Schematic representation of the *tfdA* gene cloned between the cauliflower mosaic virus 35S promoter and the nopaline synthase (*nos*) 3′ polyadenylation signal of pJ35SN and inserted into the binary vector plasmid pGA470. Km, neomycin phosphotransferase gene which mediates kanamycin resistance in transformed plants; Ap and Tc, bacterial genes for ampicillin and tetracycline resistance, respectively. B_L and B_R, left and right borders of the T-DNA, respectively

Fig. 5. Transgenic tobacco plants expressing the 2,4-D resistance gene (left) and wildtype Wisconsin 38 tobacco plants (right). (A) Plants were sprayed with 300 mg/l of 2,4-D (isopropyl ester). (B) Seed was germinated on media containing 0.4 mg/l of 2,4-D (Lyon et al., 1989)

the gene dosage, and it is possible to follow the segregation of the resistance trait by plating the seed on different concentrations of the herbicide. This provides a useful system for the selection of progeny carrying multiple copies of *tfdA* and could be useful in tracking the inheritance of other non-selectable but agronomically-valuable genes. The seed germination assay also offers a simple way to test new gene constructs for improved expression and to examine the ability of the *tfdA* gene product to detoxify other closely related herbicides. Such assays have confirmed the tight specificity of the 2,4-D monooxygenase, with activity being restricted to 2,4-dichlorophenoxyacetic acid (2,4-D), 2,4-dichlorophenoxybutyric acid (2,4-DB), 4-chloro-2-methylphenoxyacetic acid (MCPA), and 4-chlorophenoxyacetic acid.

D. Other Herbicide Candidates

A great deal of effort is being expended to develop a bacterial detoxification strategy for the broad-spectrum herbicide glyphosate (Malik et al., 1989). One research group has characterized the C-P lyase responsible for degrading glyphosate (Murata et al., 1989), however, it appears that this may not be an easy system for engineering into plants as the reaction requires the participation of at least two gene products. Surprisingly, little is known about the enzymology or genetics of the alternative AMPA pathway for the degradation of this herbicide, despite the fact that this is widespread amongst microorganisms.

Two other herbicide families for which it may be possible to engineer detoxification enzymes in plants are the diphenyl ethers and the uracils. The diphenyl ethers have a reasonably broad spectrum of activity which could be modified to advantage, and the gene for a bacterial degradative enzyme has recently been cloned (Liaw and Srinivasan, 1989). The uracil herbicides such as bromacil are currently only recommended for use on non-cropland, however, the localization of degradative functions to a bacterial plasmid may make it possible to obtain an enzyme that mediates selective resistance (Chaudhry and Cortez, 1988). As outlined in Sect. IV, bacterial degradative pathways exist for many other herbicides, however, little is known about the genetics of these processes, and a great deal of research is required before resistance to these compounds can be engineered into plants. It would be worthwhile to spend some time to determine what other crop/herbicide combinations could be of benefit to agriculture and then concentrate on characterizing and developing these systems.

VI. Conclusion

The generation of transgenic plants resistant to herbicides promises to be one of the most successful applications of recombinant DNA in agriculture. Forecasts that this technology will revolutionize weed control are unlikely to be fulfilled, but substantial gains will be made in the development of control strategies for problem weeds that are not affected by existing herbicides, through the use of lower quantities of more effective and environmentally acceptable herbicides, and in the reduced soil degradation resulting from increased applicability of minimum tillage practices.

Detoxification of the herbicide is a particularly attractive method for the achievement of this goal, and a handful of projects have shown that this strategy can work quite effectively. The detoxification enzymes chosen so far, have been the products of single genes which are relatively simple to manipulate, but future engineering efforts may be much more adventurous entailing the cloning of multi-component pathways. Under these circumstances, it will be important to ensure that there is no requirement for complex cofactors for the reactions to proceed, or any cross-specificity of the detoxification enzymes for endogenous plant compounds. The question of what happens to the detoxified herbicide left in the plant, particularly if the crop is to be consumed by humans, is something that will need to be addressed in each individual situation. If we accept the general principle of spraying crops with herbicides, however, there is no reason to single-out transgenic plants as a special case, and analysis of the fate of metabolites should be carried-out according to existing chemical residue regulations.

Some other concerns that are raised about transgenic herbicide resistant crops include the possibility that the crop itself will become a weedy species, that there could be lateral transfer of the resistance trait to weedy relatives and that sustained selection with herbicides will encourage the evolution of resistant weeds. It is unlikely that the crop possesses the necessary characteristics to become a weed in the true sense of the word (Keeler, 1989), however, there will certainly be an opportunity for these plants to act as volunteer weeds in rotation programmes and care will need to be exercised in the design of such systems. The transfer of resistance genes from the crop to closely related weeds is also a risk, particularly in centres of origin where weedy relatives may be endemic, but information is currently lacking on the likelihood of inter- and intraspecific gene transfer. Finally, the success of any one herbicide resistance strategy could ultimately lead to a high degree of genetic vulnerability if the development of resistant weeds is accelerated by increased herbicide usage. This is a more probable outcome if, as has generally been the case, the resistance traits that are being engineered are for herbicides with single targets which can readily mutate to give resistance.

Even when the scientific questions have been dealt with, there remain many obstacles along the road to the widespread application of transgenic herbicide resistant crops. Government regulatory approval, the minefield of proprietary protection and the public perception of genetically engineered crops are issues that must be resolved in a positive way. If analyzed in a rational manner, I believe that the benefits to be gained from the use of transgenic crops as an integral component in environmentally sustainable weed control strategies will prove to be irresistable.

Acknowledgments

I am grateful to my colleagues at the CSIRO Division of Plant Industry for their ideas and helpful suggestions and to Danny Llewellyn, John Huppatz, and Colin Jenkins for critical reading of the manuscript. The original work reported here was supported by Cotton Seed Distributors Pty. Ltd. and the Australian Cotton Research Council.

VII. References

Amy PS, Schulke JW, Frazier LM, Seidler RJ (1985) Characterization of aquatic bacteria and cloning of genes specifying partial degradation of 2,4-dichlorophenoxyacetic acid. Appl Environ Microbiol 49: 1237–1245

Anderson JPE (1984) Herbicide degradation in soil: influence of microbial biomass. Soil Biol Biochem 16: 483–489

Anderson JJ, Dulka JJ (1985) Environmental fate of sulfometuron methyl in aerobic soils. J Agricult Food Chem 33: 596–602

Attaway HH, Camper ND, Paynter MJB (1982) Anaerobic microbial degradation of diuron by pond sediment. Pestic Biochem Physiol 17: 96–101

Balthazor TM, Hallas LE (1986) Glyphosate-degrading microorganisms from industrial activated sludge. Appl Environ Microbiol 51: 432–434

Bartsch K, Tebbe CC (1989) Initial steps in the degradation of phosphinothricin (glufosinate) by soil bacteria. Appl Environ Microbiol 55: 711–716

Behki RM, Khan SU (1986) Degradation of atrazine by *Pseudomonas*: N-dealkylation and dehalogenation of atrazine and its metabolites. J Agricult Food Chem 34: 746–749

Bollag J-M, McGahen LL, Minard RD, Liu S-Y, (1986) Bioconversion of alachlor in an anaerobic stream sediment. Chemosphere 15: 153–162

Botterman J, Leemans J (1988) Engineering of herbicide resistance in plants. Biotechnol Genet Engineer Rev 6: 321–340

Carr RJG, Bilton RF, Atkinson T (1985) Mechanism of biodegradation of paraquat by *Lipomyces starkeyi*. Appl Environ Microbiol 49: 1290–1294

Chaudhry GR, Cortez L (1988) Degradation of bromacil by a *Pseudomonas* sp. Appl Environ Microbiol 54: 2203–2207

Chaudhry GR, Huang GH (1988) Isolation and characterization of a new plasmid from a *Flavobacterium* sp. which carries the genes for degradation of 2,4-dichlorophenoxyacetate. J Bacteriol 170: 3897–3902

Cheung AY, Bogorad L, Van Montagu M, Schell J (1988) Relocating a gene for herbicide tolerance: a chloroplast gene is converted to a nuclear gene. Proc Natl Acad Sci USA 85: 391–395

Comai L, Facciotti D, Hiatt WR, Thompson G, Rose RE, Stalker DM (1985) Expression in plants of a mutant *aroA* gene from *Salmonella typhimurium* confers tolerance to glyphosate. Nature 317: 741–744

Comai L, Stalker D (1986) Mechanism of action of herbicides and their molecular manipulation. Oxford Surv Plant Mol Cell Biol 3: 166–195

Comballack JH (1989) The importance of weeds and the advantages and disadvantages of herbicide use. Plant Protect 4: 14–32

Cook AM, Hutter R (1982) Ametryne and Prometryne as sulfur sources for bacteria. Appl Environ Microbiol 43: 781–786

Cook AM, Grossenbacher H, Hutter R (1983) Isolation and cultivation of microbes with biodegradative potential. Experientia 39: 1191–1198

De Block M, Botterman J, Vandewiele M, Dockx J, Thoen C, Gossele V, Rao Movva N, Thompson C, Van Montagu M, Leemans J (1987) Engineering herbicide resistance in plants by expression of a detoxifying enzyme. EMBO J 6: 2513–2518

De Greef W, Delon R, De Block M, Leemans J, Botterman J (1989) Evaluation of herbicide resistance in transgenic crops under field conditions. Bio/Technology 7: 61–64

della-Cioppa G, Bauer SC, Taylor ML, Rochester DE, Klein BK, Shah DM, Fraley RT, Kishore GM, (1987) Targeting a herbicide-resistance enzyme from *Escherichia coli* to chloroplasts of higher plants. Bio/Technology 5: 579–584

Don RH, Pemberton JM (1981) Properties of six pesticide degradation plasmids isolated from *Alcaligenes paradoxus* and *Alcaligenes eutrophus.* J Bacteriol 145: 681–686

Don RH, Weightman AJ, Knackmuss H-J Timmis KN (1985) Transposon mutagenesis and cloning analysis of the pathways for degradation of 2,4-dichlorophenoxyacetic acid and 3-chlorobenzoate in *Alcaligenes eutrophus* JMP134 (pJP4). J Bacteriol 161: 85–90

Donn G, Knipe B, Malvoisin P, Eckes P (1990) Field evaluation of glufosinate tolerant crops bearing a modified PPT-acetyltransferase from *Streptomyces viridochromogenes.* J Cell Biochem [Suppl 14E]: 298

Eckes P, Wengemayer F (1987) Overproduction of glutamine synthetase in transgenic plants. In: Regulation of plant gene expression. 29th Harden Conf Prog Abstr, Wye College, Ashford, UK

Fang SC (1969) Thiolcarbamates. In: Kearney PC, Kaufman DD (eds) Degradation of herbicides. Marcel Dekker, New York, pp 147–164

Gardiner JA, Rhodes RC, Adams JB, Soboczenski EJ (1969) Synthesis and studies with 2-C^{14}-labeled bromacil and terbacil. J Agricult Food Chem 17: 980–986

Gasser CS, Fraley RT (1989) Genetically engineering plants for crop improvement. Science 244: 1293–1299

Geissbuhler H (1969) The substituted ureas. In: Kearney PC, Kaufman DD (eds) Degradation of herbicides. Marcel Dekker, New York, pp 79–111

Ghosal D, You I-S, Chatterjee DK, Chakrabarty AM (1985) Microbial degradation of halogenated compounds. Science 228: 135–142

Giardina MC, Giardi MT, Filacchioni G (1982) Atrazine metabolism by *Nocardia*: elucidation of initial pathway and synthesis of potential metabolites. Agricult Biol Chem 46: 1439–1445

Goss JR, Mazur BJ (1989) A kaleidoscopic view of crop herbicide resistance. Proc Western Soc Weed Sci 42: 17–28

Haughn GW, Smith J, Mazur B, Somerville C (1988) Transformation with a mutant *Arabidopsis* acetolactate synthase gene renders tobacco resistant to sulfonylurea herbicides. Mol Gen Genet 211: 266–271

Huppatz JL (1990) Essential amino acid biosynthesis provides multiple targets for selective herbicides. In: Casida JE (ed) Pesticides and alternatives: innovative chemical and biological approaches to pest control. Elsevier, Amsterdam, pp 563–572

Jacob GS, Garbow JR, Hallas LE, Kimack NM, Kishore GM, Schaefer J (1988) Metabolism of glyphosate in *Pseudomonas* sp. strain LBr. Appl Environ Microbiol 54: 2953–2958

Kaufman DD, Blake J (1973) Microbial degradation of several acetamide, acylanilide, carbamate, toluidine and urea pesticides. Soil Biol Biochem 5: 297–308

Keeler KH (1989) Can genetically engineered crops become weeds? Bio/Technology 7: 1134–1139

Kellogg ST, Chatterjee DK, Chakrabarty AM (1981) Plasmid-assisted molecular breeding: new technique for enhanced biodegradation of persistent toxic chemicals. Science 214: 1133–1135

Kilbane JJ, Chatterjee DK, Karns JS, Kellogg ST, Chakrabarty AM (1982) Biodegradation of 2,4,5-trichlorophenoxyacetic acid by a pure culture of *Pseudomonas cepacia*. Appl Environ Microbiol 44: 72–78

Kilpi S (1980) Degradation of some phenoxy acid herbicides by mixed cultures of bacteria isolated from soil treated with 2-(2-methyl-4-chloro)phenoxypropionic acid. Microbiol Ecol 6: 261–270

Knowles CO, Benezet HJ (1981) Microbial degradation of the carbamate pesticides desmedipham, phenmedipham, promecarb and propamocarb. Bull Environ Contamin Toxicol 27: 529–533

Knuesli E, Berrer D, Dupuis G, Esser H (1969) *s*-Triazines. In: Kearney PC, Kaufman DD (eds) Degradation of herbicides. Marcel Dekker, New York, pp 51–78

Krause A, Hancock G, Minard RD, Freyer AJ, Honeycutt RC, LeBaron HM, Paulson DL, Liu S-Y, Bollag J-M (1985) Microbial transformation of the herbicide metolachlor by a soil actinomycete. J Agricult Food Chem 33: 584–589

Lappin HM, Greaves MP, Slater JH (1985) Degradation of the herbicide mecoprop [2-(2-methyl-4-chlorphenoxy)propionic acid] by a synergistic microbial community. Appl Environ Microbil 49: 429–433

Lee A (1984) EPTC (S-ethyl N,N dipropylthiocarbamate)-degrading microorganisms isolated from a soil previously exposed to EPTC. Soil Biol Biochem 16: 529–531

Liaw HJ, Srinivasan VR (1989) Cloning and expression of an *Erwinia* sp. gene encoding diphenyl ether cleavage in *Escherichia coli*. Appl Environ Microbiol 55: 2220–2225

Liu T, Chapman PJ (1984) Purification and properties of a plasmid-encoded, 2,4-dichlorophenol hydroxylase. FEBS Lett 173: 314–318

Lyon BR, Llewellyn DJ, Huppatz JL, Dennis ES, Peacock WJ (1989) Expression of a bacterial gene in transgenic tobacco plants confers resistance to the herbicide 2,4-dichlorophenoxyacetic acid. Plant Mol Biol 13: 533–540

MacRae IC (1989) Microbial metabolism of pesticides and structurally related compounds. Rev Environ Contamin Toxicol 109: 1–87

Malik J, Barry G, Kishore G (1989) The herbicide glyphosate. BioFactors 2: 17–25

Marty JL, Vouges J (1987) Purification and properties of a phenylcarbamate herbicide degrading enzyme of *Pseudomonas alcaligenes* isolated from soil. Agricult Biol Chem 51: 3287–3294

Mazur BJ, Falco SC (1989) The development of herbicide resistant crops. Annu Rev Plant Physiol Plant Mol Biol 40: 441–470

Mazur BJ, Chui C-F, Smith JK (1987) Isolation and characterization of plant genes coding for acetolactate synthase, the target enzyme for two classes of herbicides. Plant Physiol 85: 1110–1117

McBride KE, Kenny JW, Stalker DM (1986) Metabolism of the herbicide bromoxynil by *Klebsiella pneumoniae* subsp. *ozaenae*. Appl Environ Microbiol 52: 325–330

McGahen LL, Tiedje JM (1978) Metabolism of two acylanilide herbicides, Antor herbicide (M-22234) and Dual (metolachlor) by the soil fungus *Chaetomium globosum*. J Agricult Food Chem 26: 414–419

Moore JK, Braymer HD, Larson AD (1983) Isolation of a *Pseudomonas* sp. which utilizes the phosphonate herbicide glyphosate. Appl Environ Microbiol 46: 316–320

Murata K, Higaki N, Kimura A (1989) A microbial carbon-phosphorus bond cleavage enzyme requires two protein components for activity. J Bacteriol 171: 4504–4506

Novick NJ, Alexander M (1985) Cometabolism of low concentrations of propachlor, alachlor, and cycloate in sewage and lake water. Appl Environ Microbiol 49: 737–743

Novick NJ, Mukherjee R, Alexander M (1986) Metabolism of alachlor and propachlor in suspensions of pretreated soils and in samples from ground water aquifers. J Agricult Food Chem 34: 721–725

O'Keefe DP, Romesser JA, Leto KJ (1988) Identification of constitutive and herbicide inducible cytochromes P-450 in *Streptomyces griseolus*. Arch Microbiol 149: 406–412

Perkins EJ, Lurquin PF (1988) Duplication of a 2,4-dichlorophenoxyacetic acid monooxygenase gene in *Alcaligenes eutrophus* JMP134 (pJP4). J Bacteriol 170: 5669–5672

Pipke R, Amrhein N (1988a) Degradation of the phosphonate herbicide glyphosate by *Arthrobacter atrocyaneus* ATCC 13752. Appl Environ Microbiol 54: 1293–1296

Pipke R, Amrhein N (1988b) Isolation and characterization of a mutant of *Arthrobacter* sp. strain GLP-1 which utilizes the herbicide glyphosate as its sole source of phosphorus and nitrogen. Appl Environ Microbiol 54: 2868–2870

Pipke R, Schulz A, Amrhein N (1987) Uptake of glyphosate by an *Arthrobacter* sp. Appl Environ Microbiol 53: 974–978

Potrykus I (1989) Gene transfer to cereals: an assessment. Tibtech 7: 269–273

Probst GW, Tepe JB (1969) Trifluralin and related compounds. In: Kearney PC, Kaufman DD (eds) Degradation of herbicides. Marcel Dekker, New York, pp 255–282

Romesser JA, O'Keefe DP (1986) Induction of cytochrome P-450-dependent sulfonylurea metabolism in *Streptomyces griseolus*. Biochem Biophys Res Comm 140: 650–659

Saxena A, Zhang R, Bollag J-M (1987) Microorganisms capable of metabolizing the herbicide metolachlor. Appl Environ Microbiol 53: 390–396

Shah DM, Horsch RB, Klee HJ, Kishore GM, Winter JA, Tumer NE, Hironake CM, Sanders PR, Gasser CS, Aykent S, Siegal NR, Rogers SG, Fraley RT (1986) Engineering herbicide tolerance in transgenic plants. Science 233: 478–481

Shah DM, Gasser CS, della-Cioppa G, Kishore GM (1988) Genetic engineering of herbicide resistance genes. In: Verma DPS, Goldberg RB (eds) Temporal and spacial regulation of plant genes. Springer, Wien New York, pp 297–309 [Dennis ES et al (eds) Plant gene research. Basic knowledge and application]

Schulz A, Wengenmayer F, Goodman HM (1990) Genetic engineering of herbicide resistance in higher plants. CRC Crit Rev Plant Sci 9: 1–15

Shinabarger DL, Braymer HD (1986) Glyphosate metabolism by *Pseudomonas* sp. strain PG2982. J Bacteriol 168: 702–707

Smith AE, Hayden BJ (1981) Relative persistence of MCPA, MCPB and mecoprop in Saskatchewan soils and the identification of MCPA in MCPB-treated soils. Weed Res 21: 179–183

Smith AE (1989) Transformation of the herbicide [^{14}C]glufosinate in soils. J Agricult Food Chem 37: 267–271

Stalker DM, McBride KE, Malyi LD (1988) Herbicide resistance in transgenic plants expressing a bacterial detoxification gene. Science 242: 419–422

Stalker DM, McBride KE (1987) Cloning and expression in *Escherichia coli* of a *Klebsiella ozaenae* plasmid-borne gene encoding a nitrilase specific for the herbicide bromoxynil. J Bacteriol 169: 955–960

Stanier RY, Palleroni , NJ, Doudoroff M (1966) The aerobic pseudomonads: a taxonomic study. J Gen Microbiol 43: 159–271

Stralka KA, Camper ND (1981) Microbial degradation of profluralin. Soil Biol Biochem 13: 33–38

Streber WR, Willmitzer L (1989) Transgenic tobacco plants expressing a bacterial detoxifying enzyme are resistant to 2,4-D. Bio/Technology 7: 811–816

Streber WR, Timmis KN, Zenk MH (1987) Analysis, cloning, and high-level expression of 2,4-dichlorophenoxyacetate monooxygenase gene *tfdA* of *Alcaligenes eutrophus* JMP134. J Bacteriol 169: 2950–2955

Tachibana K, Watanabe T, Sekizawa Y, Takematsu T (1986) Accumulation of ammonia in plants treated with bialaphos. J Pestic Sci 11: 33–37

Talbot HW, Johnson LM, Munnecke DM (1984) Glyphosate utilization by *Pseudomonas* sp. and *Alcaligenes* sp. isolated from environmental sources. Curr Microbiol 10: 255–260

Tam AC, Behki RM, Khan SU (1987) Isolation and characterization of an *s*-ethyl-*N,N*-dipropylthiocarbamate-degrading *Arthrobacter* strain and evidence for plasmid-associated *s*-ethyl-*N,N*-dipropylthiocarbamate degradation. Appl Environ Microbiol 53: 1088–1093

Thompson CJ, Rao Movva N, Tizard R, Davies JE, Lauwereys M, Botterman J (1987) Characterization of the herbicide-resistance gene *bar* from *Streptomyces hygroscopicus*. EMBO J 6: 2519–2523

Tiedje JM, Hagedorn ML (1975) Degradation of alachlor by a soil fungus, *Chaetomium globosum*. J Agricult Food Chem 23: 77–81

Vega D, Bastide J, Coste C (1985) Isolation from soil and growth characteristics of a CIPC-degrading strain of *Pseudomonas cepacia*. Soil Biol Biochem 17: 541–545

Wang Y-S, Subba-Rao RV, Alexander M (1984) Effect of substrate concentration and organic and inorganic compounds on the occurrence and rate of mineralization and cometabolism. Appl Environ Microbiol 47: 1195–1200

Wright SJL, Maule A (1982) Transformation of the herbicides propanil and chloropropham by micro-algae. Pestic Sci 13: 253–256

Zeyer J, Kearney PC (1982) Microbial metabolism of propanil and 3,4-dichloroaniline. Pestic Biochem Physiol 17: 224–231

Zeyer J, Kearney PC (1983) Microbial dealkylation of trifluralin in pure culture. Pestic Biochem Physiol 20: 10–18

Chapter 6

The Manipulation of Plant Gene Expression Using Antisense RNA

Wolfgang Schuch

ICI Seeds, Plant Biotechnology Section, Jealott's Hill Research Station,
Bracknell, Berks. RG12 6EY, U.K.

Contents

I. Introduction

Gene cloning techniques have allowed us to analyse in considerable detail the structure of plant genes. The availability of probes for plant genes has opened the way for detailed studies on the temporal and spatial expression of plant genes during the development of different plant organs and tissues (Kuhlemeier et al., 1987). Functional analysis of isolated plant genes was made possible through the development of gene transfer techniques (Weising et al., 1988). This has led to the detailed description of expression patterns exhibited by gene promoter sequences, and to the delineation of

DNA sequences which are required for the temporal and spatial control of gene expression.

There is also now an ever increasing abundance of characterised cDNA clones from a large number of different plants, tissues and stages of development which have been isolated and studied. These are obviously outnumbered by the even larger number of clones present in cDNA or genomic libraries which have not been identified by biochemical function, expression levels, and tissue and cell specificity. Therefore, one of the challenges for the future will lie in the development of techniques which will permit the functional characterisation of these clones.

In several plant systems mutants are available which are being used in the functional identification and analysis of plant mRNAs and genes (Thomas and Grierson, 1987). In most cases, these mutants are characterised by the aberrant expression of specific genes, either the absence of expression or the reduction of gene expression leading to a specific phenotype, which was either phenotypically scorable or biochemically measurable.

It was only recently, that methods have been developed for the generation of novel plant mutants. Several approaches have been proposed in the literature such as antisense RNA to downregulate the expression of genes (Izant and Weintraub, 1984), homologous recombination to knock out the expression of genes (Smithies et al., 1985), or dominant negative mutations to inhibit the formation of functional enzyme complexes (Herskowitz, 1987). The most successful method developed over the past two years uses the expression of antisense RNA for the inhibition of specific genes. This approach is based on the observation that in several naturally occurring situations RNA:RNA interactions control the expression of genes (Green et al., 1986; Pines and Inouye, 1986; Simons and Kleckner, 1988). These RNA–RNA interactions depend on the base pairing of sense and antisense RNA molecules. The generation of double stranded RNA molecules then leads to the removal of the functional RNA leading to control of gene expression.

The effect of antisense RNA has been studied in vitro (Paterson et al., 1977; Hastie et al., 1978). This work has then been extended to study the control of gene expression in transient systems (Harland and Weintraub, 1985; Melton, 1985). Recently, these techniques have also been applied to the study of plant gene expression and development (for references see below). These experiments have led to the generation of novel plant mutants which exhibit reduced expression of endogenous plant genes and altered phenotypes. The possibility now exists to use this method to identify the biochemical function of plant genes whose function was previously unknown.

This subject area has been reviewed in detail by others (van der Krol et al., 1988b). The authors have given a full review of the natural occurrence of antisense RNA control in prokaryotic systems and have also described in detail the use of antisense oligonucleotides. In addition, several papers have occurred in the literature which describe the use of antisense RNA to control viral infection of plants (Beachy et al., 1987; Loesch-Fries et al., 1987; Hemenway et al., 1988; Rezaian et al., 1988). In this article I will only review experiments describing the control of expression of genes in transformed plants.

II. Examples of the Control of Plant Gene Expression Using Antisense RNA

A. Control of Expression in a Transient Assay

Ecker and Davis (1986) were the first to demonstrate successful down-regulation of gene expression in plant cells using antisense RNA. A series of vectors were constructed using the nopaline synthase (*nos*), CaMV 35S and 'PAL' promoters to drive the expression of either the sense or antisense RNA encoding the complete bacterial chloramphenicol acetyl transferase (CAT) gene. Sense and antisense vectors were then electroporated together into carrot cell protoplasts and the level of CAT expression was measured. The ratios of sense to antisense plasmids in the electroporation experiments were varied between 1:1 to 1:10 to 1:100. In the best experiments a reduction in CAT activity of 95% was observed. This was achieved using the *nos* promoter to drive the expression of antisense RNA. The ratio of sense to antisense plasmid was 1:100. When a *nos* polyadenylation signal was included in the antisense vector, a similar great reduction of CAT activity was observed. However, now the ratio required was 1:50. In further experiments using the different promoter combinations, the reduction in CAT expression was correlated with the strength of the promoter driving the antisense gene.

B. Control of Expression in Transgenic Plants

1. Control of 'Model' Genes in Tobacco

Inhibition of Nopaline Synthase (nos). Rothstein et al. (1987) were the first to show the stable and heritable inhibition of a model gene introduced into tobacco. Tobacco plants transformed with the *Agrobacterium* transformation vector pTiT37 containing a nopaline synthase gene (*nos*) were retransformed with a vector which contained the first 860 bases (approximately two thirds) of the *nos* gene in an antisense orientation. The promoter used to drive the expression of this gene was the CaMV 35S promoter. Transformed plants were regenerated, and grown in soil in the glasshouse before being analysed. Plants transformed with the binary transformation vector alone were used as controls.

Considerable variation in *nos* activity was observed in plants of different ages. This variation was observed in both plants transformed with the control as well as the antisense vector. However, a reduction in *nos* activity between 10 and 50 fold was also detectable. The reduction was seen in plants grown in tissue culture pots as well as those grown in the glasshouse. RNAse protection experiments indicated that the levels of *nos* RNA was considerably reduced in the antisense plants when compared to controls (approximately 8–10 fold). The levels of antisense RNA found were approximately equal to the level of *nos* RNA in control plants. The CaMV 35S promoter was used to drive the expression of the *nos* antisense RNA. The endogenous *nos* gene was driven by its own promoter. From other experiments it was expected that the levels of antisense RNA would be 30 fold higher than the *nos* sense RNA levels. The interpretation of this discrepancy was that the antisense mRNA was apparently degraded faster than *nos* RNA. The decrease in the level of *nos* activity could be accounted for by the decrease in the level of *nos* mRNA. These authors point out that the steady state level of antisense RNA was approximately the same as that of sense RNA in control plants. Therefore no great excess of antisense RNA was required in these experiments to achieve the reduction in *nos* activity. A genetic analysis was also performed which indicated that the reduced *nos* phenotype was stably inherited. These experiments demonstrated co-segregation of the hygromycin resistance marker with the reduced *nos* phenotype.

In a more detailed study, others (Sandler et al., 1988) performed similar experiments. A transgenic tobacco plant, JAT20, was generated by transformation of *Nicotiana tabaccum* Wisconsin protoplasts with pGV3850. This yielded plants which were *nos* +. A series of antisense vectors were constructed using the CAB 22R promoter from petunia including the 5′ untranslated region of the gene. The 3′ untranslated region was provided by the *ocs* gene. The vectors generated cover a range of the *nos* gene from the full-length construct pSJS219 (1565 bases) to pSJS225 (260 bases). As the CAB promoter drives the expression of foreign proteins and mRNA levels to approximately 20 to 30 times those of the *nos* promoter, considerable excess of antisense RNA can be generated in these plants. For each construct between 7 and 15 plants were regenerated. This allowed these authors to perform a detailed statistical analysis which could cope with the variation between individual transformants. This variation was observed in tissue- and developmental-stage specific expression of *nos* in JAT20. For the analysis of antisense plants three different leaves were harvested from plants at different stages of growth. The samples were then pooled and analysed using two-way analysis of variance. This indicated that there was considerable and significant variation in *nos* activity in plant material harvested at different dates during the growing season. The samples from each construct were pooled for each harvest date, and the data pooled for all harvest dates from each constructs. They were then compared with those of control populations (controls being untransformed plants or plants transformed with a vector lacking the antisense gene). This analysis indicated that significant differences existed between some of these populations: the controls and those antisense constructs which covered the 3′ half of the *nos* gene. These inhibitory constructs ranged from 261 bases to approximately 900 bases. These data indicated that specific sequences may give a more effective control of gene expression when compared to others derived from the same gene. In none of the experiments was *nos* activity completely inhibited.

Inhibition of Chloramphenicol Acetyl Transferase (CAT). Delauney et al. (1988) used a different approach to demonstrate antisense RNA control in tobacco plants. These authors generated transgenic tobacco plants expressing the bacterial CAT gene. The CAT gene was expressed from the CaMV 19S promoter. Plants were identified which carried one (P41) and several (S13) CAT genes at a single locus. The levels of CAT expression of S13 were higher than those observed in P41. These two plants were then retransformed with a vector in which the CAT sequences in antisense orientation were placed under the transcriptional control of the *rbcS* promoter isolated from *Nicotiana plumbaginifolia* and the *nos* terminator. Twentyfive regenerated plants were analysed and none were found to have any reduction in CAT activity. The interpretation of these results was that the antisense RNA may be unstable and therefore not available for inhibition of sense CAT RNA. An alternative explanation is that the expression of sense and antisense RNA did not take place in the same cells. Therefore, another strategy was developed. The CAT antisense sequences were linked to a selectable marker gene (hygromycin resistance) in order to produce a bifunctional mRNA. The expression of the hygromycin-resistance-gene-CAT-antisense construct was driven by the CaMV 35S promoter. The reasoning behind this was that the CaMV 35S promoter is capable of a ten fold greater expression level that the CaMV 19S promoter. The vector which contained this cassette was modified in such a way that the kanamycin resistance gene was removed. This permitted the selection of transformed plants containing the original CAT sequences by selection on kanamycin, and the selection of the antisense sequences using hygromycin selection. Two vectors were constructed, containing either the complete CAT antisense gene or only the 172 5′ nucleotides. Of 32 plants tested from the progeny of a parent (P41) plant transformed with the complete CAT antisense gene, 9 were partially inhibited (~40% of control levels), and 6 were completely inhibited. When the 5′ region was used in the construction of the antisense vector, 10 plants out of 18 showed a reduction, but none gave complete inhibition of CAT expression. When the parent plant used for retransformation was exhibiting high levels of CAT activity (S31; giving 5 times higher levels of CAT expression than P41), only 2 out of 22 plants analysed showed complete inhibition of CAT activity. Four plants showed intermediate levels of CAT activity. The shorter construct did not give as effective control in this case (5/15 plants) as had been observed in the previous experiment.

RNA analysis of those plants in which CAT activity had been completely inhibited showed high levels of expression of the fused transcript. In the retransformed P41 plants levels between 2 to 20 times the levels of CAT sense mRNA were observed. In the S13 transformants which were CAT^- the levels of CAT antisense RNA were approximately the same as those found in control plants. Plants which showed the highest levels of CAT expression also showed the lowest levels of antisense RNA present. These plants also showed detectable levels of sense CAT RNA which were comparable with normal CAT RNA found in the parent plants. In those plants which were selected on high levels of hygromycin but did not show any reduction in CAT activity, no evidence of the bifunctional mRNA was detected. All plants retransformed with a construct lacking CAT antisense DNA showed normal levels of CAT activity.

These CAT enzyme and RNA data were further supported by DNA analysis of the transformed plants. This indicated that the reduction in CAT activity was not due to a rearrangement of the CAT DNA sequences.

Inhibition of β-Glucuronidase (GUS). Recently another example of the inhibition of a bacterial marker gene introduced into plant cells was provided (Robert et al., 1989). These authors asked the question whether antisense RNA will effectively control the expression of a gene when both genes are expressed from the same promoter. Transgenic tobacco plants were generated by transformation with the pBI121.1 vector. In this vector the bacterial GUS gene is driven by the CaMV 35S promoter. A transformant was identified which contained a single GUS gene and exhibited high levels of GUS activity (TTR-48, GUS^+). This plant was then retransformed with a vector which contained an antisense GUS gene driven by the CaMV 35S promoter. This vector also permitted the selection of newly transformed plants by virtue of the presence of the hygromycin resistance gene. This vector was named PAL 1302. Considerable variation in GUS activity was observed in those plants retransformed with PAL 1302. Furthermore, a developmental stage dependent variation of the reduction of GUS expression was observed. Five plants (TTR1–TTR5) gave the greatest reduction in GUS activity at all stages of development. The highest level of reduction in GUS activity was approximately 90%. In plants which were retransformed with the vector without the antisense construct, no reduction in GUS activity was apparent. The reduction in GUS enzyme activity was closely correlated with the reduction in GUS protein as measured by Western blotting. Southern hybridisation confirmed the presence of the

GUS sense and antisense genes in these plants. Northern hybridisation was then used to test for the presence of GUS sense and antisense RNAs. GUS sense RNA was observed, however only at greatly reduced levels.

Inhibition of Phosphinothricin Acetyl Transferase (PAT). Phosphinothricin acetyl transferase (PAT) from *Streptomyces* has been introduced into tobacco plants. This gene confers resistance to bialaphos, a potent herbicide. The level of expression of PAT protein achieved in transgenic tobacco plants is approximately 0.2% of total soluble protein. The protein is also expressed in leaf protoplasts. Cornelissen and Vandewiele (1989) generated a plant transformation vector in which the CaMV 35S promoter was used to drive the expression of *bar* antisense RNA. The complete *bar* coding region was used in the antisense vector. This vector was transferred to tobacco SR1, generating a tobacco line SR1 (T-GSC1) in which full-length *bar* antisense RNA was detected. Protoplasts of this line as well as the untransformed control (SR1) were used in electroporation experiments, in which a hybrid construct consisting of the *bar* gene and a CAT gene was introduced. CAT and PAT activity were determined. This experiment led to the conclusion that PAT activity was specifically reduced in this transient assay up to 90 min after electroporation whereas CAT activity was not affected.

Transgenic tobacco plants were generated by transformation of SR1 with a construct carrying the *bar* gene expressed from the TR2′ promoter and a hygromycin resistance marker gene. A transformed plant was identified which contained two copies of this construct (SR1–GSFR166). This plant was then retransformed with the antisense vector. Six independent lines were analysed for PAT activity. Two lines were identified which showed a reduction in PAT activity of approximately 90%. One plant was chosen for further analysis [SR1(T-GSFR166, T-GSC1)]. Leaf protoplasts from this and the parent line were taken and the rate of de novo PAT protein synthesis determined using in vivo labelling techniques combined with immunoprecipitation. This indicated that the levels of PAT protein in the antisense containing plant were 13 fold less than in the controls.

The levels of hpt, *bar* and anti-*bar* mRNA were determined using slot blot analysis. This analysis showed that the levels of *bar* mRNA were reduced in anti-*bar* containing cells whereas the levels of hpt transcripts were at a similar level. However, the reduction in *bar* mRNA (four fold) were different from the reduction in *bar* protein observed (13 fold). The authors concluded that in addition to a reduction of *bar* mRNA an effect on translatability of the *bar* mRNA was also seen.

2. Control of Endogenous Genes in Tobacco, Petunia, and Tomato

Inhibition of Ribulose Bisphosphate Carboxylase (rbc). Rodermel et al. (1988) used antisense RNA as an approach to study regulation of the assembly of the small and large subunits of RuBisCo in tobacco. The antisense RNA expression vector, pTASS contained almost the complete RuBisCo small subunit mRNA covering 22 bases 5′ of the translation initiation region to 300 bases of the coding region of the mRNA. This was cloned into an expression vector which uses the CaMV 35S promoter and the CaMV 35S 3′ end. The *rbcS* cDNA was isolated from tobacco species *Nicotiana sylvestris.* This vector was transformed into tobacco SR1. Sequence homology between the *rbcS* genes from these different tobacco species is greater than 99.7%. Five transformants were recovered in which a single antisense gene was present, whereas in the remaining plants several antisense genes had been inserted. Considerable variation was detected in the levels of *rbcS* transcript in these plants. In plants containing a single antisense gene construct the level was reduced by approximately 80%, whereas plants containing multiple antisense gene insertions showed a reduction of 90%. Antisense RNA transcripts were detected using strand specific probes only in plant 5, which contained several antisense gene inserts. In addition to a reduction of *rbcS* mRNA, *rbcS* protein was also reduced. Thus these plants represent novel mutants in which the coordinate expression and assembly of RuBisCo subunits can be studied, as natural mutants in which the co-ordination of RuBisCo assembly is affected are not available. It was demonstrated that the normal stoichiometry of the subunit pool size was not altered in the antisense plants. The accumulation of holoenzyme was correlated with the level of *rbcS* mRNA. On the other hand *rbcL* mRNA levels were not correlated with the accumulation of RuBisCo. This analysis indicated that transcriptional and post-transcriptional effects are implicated in the assembly of RuBisCo.

One transformant, plant 5, was selfed and the progeny analysed. Progeny plants showed different growth rates depending on the number of antisense genes present. Plants containing no antisense gene gave normal growth, those with intermediate numbers of antisense genes gave intermediate growth, and those containing the largest numbers of antisense genes, were severely stunted.

Inhibition of Chalcone Synthase (CHS). Van der Krol et al. (1988a) used the extensive knowledge of the biochemistry, genetics, and molecular biology of the flavanoid pathway in petunia to control the expression of chalcone synthase. CHS is a key enzyme in the synthesis of flavanoids which lead to the great variation in flower pigmentation in many plants. An antisense vector was designed which utilised the CaMV 35S promoter to drive the expression of CHS cDNA. The cDNA insert used in the antisense vector consisted of the complete coding region including the 3′ untranslated region of the CHS mRNA. This vector, VIP104, was introduced into petunia VR hybrid. The resultant transformants were inspected for changes in flower pigmentation.

The plants fell into three categories: the first, in which no alteration in pigmentation was observed (12 out of 25 plants). The second, in which a reduction in pigmentation in the corolla of the flower was observed (8 out of 25 plants). In these plants the tube of the flower was coloured as normal. In the third class, the alteration in pigmentation started in the tube, and extended to the corolla, yielding almost completely colourless flowers. In some instances, flowers were generated in which novel patterns of pigmentation were observed. Biochemical analysis indicated that flavanoid biosynthesis had been affected, as TLC analysis of white flower tissue extracts demonstrated the absence of pigments. It has also been demonstrated, that the enzymes following CHS in the pathway were not affected. This has been achieved by biochemical feeding experiments, which restore pigment formation.

Southern hybridisation was used to prove the presence of the antisense genes in these plants. Northern hybridisation on selected plants of each of the above classes demonstrated the presence of antisense transcripts. However, two transcripts were seen which can be accounted for if different polyadenylation sites in the vector are being used. This analysis indicated that antisense RNA was detected in the target tissues. Expression of antisense RNA in the leaves did not correlate with the levels of reduction in pigmentation observed in the flower. Plants of the three different classes described above, showed similar levels of antisense RNA in the leaves. Plants which show high levels of steady state antisense RNA may be found in the class in which no inhibition of pigment formation is seen. On the other hand, plants in which low levels of steady state antisense RNA is observed in the leaf, show the greatest level of inhibition of floral pigmentation.

A number of plants containing the antisense genes were backcrossed to the two parents of the VR hybrid (V23 and R51). Complete co-segregation of the antisense phenotype with the antisense gene was observed.

The petunia antisense gene was also used to transform tobacco plants. As quoted in van der Krol et al. (1988), the inhibition of tobacco CHS was also observed, although the sequence divergence between the tobacco and petunia CHS genes has been estimated to be around 20%.

The CHS gene itself has also been used in the construction of antisense vectors (J.N.M. Mol, pers. comm.). The results from these experiments indicate that only fragments covering the 3′ end of the CHS gene gave effective reduction in CHS activity. As this region of the gene does not cover the intron, it was concluded that intron excision is not a possible target for antisense RNA function.

Inhibition of Polygalacturonase (PG). We have designed experiments aimed at controlling the expression of polygalacturonase (PG), a major cell wall enzyme in tomato fruit. There has been considerable interest in this enzyme, as it has been postulated to be responsible for fruit softening

(Hobson, 1964; Hobson, 1965; Brady et al., 1982) through changes in the pectin fraction of the tomato cell wall (Tucker and Grierson, 1983; Crookes and Grierson, 1983). PG is synthesised de novo at the onset of fruit ripening (Tucker and Grierson, 1982). Studies on the control of PG gene expression have demonstrated that this increase is due to an increase in transcription of the PG gene (Grierson et al., 1986; DellaPenna et al., 1986). At its maximal level PG mRNA represents approximately 1% of the total mRNA mass of the fruit. PG protein represents approximately 15% of the cell wall protein of the ripe tomato fruit. The PG gene is expressed only during tomato fruit ripening. The PG cDNA and gene have been isolated in our laboratory (Grierson et al., 1986; Birol et al., 1988) and by others (Sheehy et al., 1987). Using the PG cDNA we have designed several experiments to downregulate PG gene expression.

We have used the CaMV 35S promoter to drive the constitutive expression of PG antisense RNA. We chose this approach as we wanted to achieve high levels of PG antisense RNA expression in the mature green fruit before the onset of PG gene expression. We have demonstrated that the CaMV 35S promoter allows expression of the bacterial CAT gene in transgenic tomatoes both in mature green (before the onset of PG gene expression) and fully red fruit (Smith et al., unpubl.).

We have chosen a 730 bp fragment derived from the 5′ end of the PG cDNA to construct an antisense RNA vector, pJR16A (Smith et al., 1988). The 730 bp fragment contains 50 bases of 5′ untranslated region, and 680 bases of coding region. The antisense gene construct was transferred to tomatoes and 13 independent transformants have been generated.

In the first experiments we described the characterisation of GR16 and GR15, two independent transformants (Smith et al., 1988). The levels of PG protein and PG enzyme activity in the cell wall preparations was reduced by 90% (GR16) and 72% (GR15) respectively. There was a correlation between the reduction in PG enzyme activity and the level of PG protein detected. In addition, the level of PG enzyme activity was reduced throughout fruit ripening. The accumulation of lycopene as an indicator of ripening, was not distinguishable in these plants from controls. Northern hybridisations demonstrated the presence of antisense RNA in both the leaves, mature green and ripe fruit RNAs. RNA samples from the ripe transformed fruit showed a reduction in the antisense as well as the sense PG mRNAs.

We have now analysed three further independent transformants (Smith et al., 1990). The levels of reduction of PG enzyme in these plants varied from 49% to 95%. The presence of the antisense gene was detected using Southern hybridisation. All plants contained one antisense gene which was found on different restriction enzyme fragments in the different plants. We have concluded, that the antisense gene had been inserted into different chromosomal locations. Thus the variation in the levels of PG enzyme activity may be due to positional effects.

We have also selfed GR16, and have analysed eleven progeny plants (Smith et al., 1990). Plants were classified by Southern hybridisation as those which were homozygous and hemizygous for the antisense gene and those in which the antisense gene has been lost through segregation. In these plants the levels of PG enzyme activity were $\sim 1\%$ (homozygous, GR105), 20% (hemizygous, GR95) and 100% (no antisense gene, GR102). Thus, a clear copy number effect in efficacy of antisense control was seen.

The expression of PG enzyme during fruit ripening in three lines representative of each genotype was studied (GR105, GR95, and GR102) (Schuch et al., 1989; Smith et al., 1990). The results indicate that plants which have lost the antisense gene have normal levels of PG. The developmental pattern of PG accumulation is identical to that of control plants. Plants hemizygous for the antisense gene have PG enzyme levels identical to those observed in the original transformant. Plants homozygous for the antisense gene have greatly reduced levels of PG enzyme activity. This low level remains the same throughout fruit ripening. In further generations of homozygous antisense plants, the levels of PG remains low, indicating that the PG antisense gene is inherited as a stable nuclear gene.

PG antisense RNA was detectable in the leaves of these plants, however, there was no correlation with gene copy number. Different size antisense RNA molecules have been detected which can be explained by premature termination/polyadenylation at sites within the antisense gene. In plants homozygous for the antisense gene, a small antisense RNA was detectable whose function is unknown.

It is noteworthy, that the timing of the increase of PG enzyme in heterozygous antisense plants is the same as that in control plants. It is only after the initial rise in PG enzyme activity that an inhibition of PG occurs. In the light of the design of the experiment (i.e., PG antisense RNA was present at high levels in the mature green fruit before the onset of PG gene expression) it is surprising that the presence of the antisense RNA was not more effective in controlling PG enzyme accumulation at the early stages of PG gene expression. When PG mRNA accumulation in the controls was maximal, a constant level of PG enzyme activity was found, indicating that control of PG gene expression was effective at this stage (Schuch et al., 1989).

We have also determined changes for several other biochemical characteristics such as lycopene accumulation, ethylene evolution, invertase activity, and more importantly, pectinesterase activity (another enzyme involved in cell wall metabolism) in these plants. None of these parameters were found to be impaired in these plants.

Thus, novel tomato mutants have been created in which PG gene expression has been specifically reduced. This is significant, as all other ripening mutants of tomato described to date have pleiotropic effects, i.e., several phenotypic and biochemical processes are affected (Grierson et al.,

1987). Thus, these mutant lines can now be used to determine the biochemical role of PG during fruit ripening, and its relevance to fruit softening. As PG is involved in pectin metabolism, we have used these mutants to measure the average molecular weight of pectins in tomato cell walls. In control plants a significant reduction in the average molecular weight of pectins during fruit ripening is observed (Seymour and Harding, 1987). Plants homozygous for the antisense gene, do not show this reduction in pectin molecular weight. Plants hemizygous for the antisense gene show an intermediate reduction in the level of pectin degradation. The overall levels of soluble pectin in antisense and control fruit is unaffected. This result clearly indicates that the major role of PG in fruit ripening is the degradation of solubilised pectin. The role of PG in the solubilisation of pectin observed at the onset of ripening is not clear. It is interesting to note, that the residual activity of PG measured in antisense fruit is primarily the PG1 isoenzyme.

Thus, it is interesting to speculate that the in vivo role of PG2 isoenzymes is related to the depolymerisation of high molecular weight soluble pectin. In order to define whether PG1 also plays a role in pectin metabolism further plants need to be generated in which PG activity is reduced even further. We are now in the process of generating double homozygous PG antisense plants containing 3 and 4 copies of the antisense gene. These plants will be used in further biochemical studies to elucidate the biochemical role of PG during fruit ripening and the pathway of PG transport into the cell wall.

We have now also grown the homozygous antisense plants (GR105) and plants which have lost the antisense gene in the selfing process under commercial glasshouse growing conditions to analyse the effect of PG antisense RNA on fruit quality measurements. Analysis of these data indicate that the reduction in PG leads to an improvement in tomato processing characteristics and storability (Schuch et al., in prep.).

We have described the isolation and characterisation of a PG promoter fragment (Bird et al., 1988) and have constructed PG antisense vectors in which the antisense RNA is expressed from the PG promoter. Transgenic plants have been generated and are presently being analysed.

Sheehy et al. (1988) have used a similar approach for the down regulation of PG. These authors have used the full-length PG cDNA for the expression of antisense RNA under the control of the CaMV 35S promoter. Ten independent transformants were analysed. RNA was extracted from immature green fruit. As PG mRNA is not expressed at this stage of fruit development, RNA samples were probed with the PG cDNA clone. mRNAs of 1.8, 2.1 and 3.6 kb were observed, all of which can be explained by the use of different polyadenylation signals. The very long antisense RNA may be due to transcriptional readthrough of the terminator in the transformation vector which was derived from transcript 7 of the T-DNA. PG antisense

RNA was also found in the leaves. One plant 1616-1 was investigated further. During ripening PG antisense RNA levels were lower than those found for PG mRNA during normal ripening. The level of PG antisense RNA remained more or less the same during ripening (with a slight decrease at later stages of ripening). However, the steady state level of PG mRNA was strongly inhibited. The levels in the fruit were determined using strand specific probes, and were found to be approximately 10% of normal levels. At 'extremely' late stages of ripening, the level of PG mRNA was also reduced. The rate of transcription of the PG gene and the PG antisense gene were measured. This analysis indicated that the antisense gene was transcribed at higher levels than the PG gene. It also indicated that the rate of PG gene transcription was not affected in transformed fruit when compared with control fruit. PG enzyme activity was reduced to $\sim 80\%$ in this plant. In other plants a reduction of between 69% and 93% was detected. There was no correlation between apparent levels of antisense RNA and reduction in PG enzyme activity. Lycopene accumulation was not affected in these plants.

Inhibition of Pectinesterase (PE). PE is another enzyme involved in the metabolism of cell walls in tomato. We have cloned a PE cDNA (Ray et al., 1988) and designed antisense RNA vectors similar to those described for PG in the previous section (Schuch et al., unpubl.). Several transgenic plants have been generated and analysed for the level of PE enzyme. We have now identified plants in which PE activity is reduced. Further work is in progress to characterise these plants in detail at the biochemical level.

Inhibition of pTOM13. We have isolated several ripening related cDNA clones from a ripe tomato cDNA library (Slater et al., 1985). One of these clones, pTOM13 has been analysed in detail by Holdsworth et al. (1987). This clone is derived from a small multigene family (Holdsworth et al., 1988). The expression of this clone has been studied in detail during tomato fruit ripening as well as wounding of tomato fruit and leaves (Smith et al., 1986; Holdsworth et al., 1988). This analysis has indicated that a close association exists between the expression of pTOM13 RNA and ethylene biosynthesis. Hamilton, Lycett and Grierson have designed antisense RNA vectors in which the expression of pTOM13 antisense RNA is driven by the CaMV 35S promoter (pers. comm.). Transformed tobacco and tomato plants have been generated and expression of pTOM13 antisense RNA studied. In transgenic tomato plants, pTOM13 antisense RNA was measured in the leaves and fruit. Upon wounding or after the onset of normal pTOM13 expression during fruit ripening, the level of pTOM13 antisense RNA is reduced. This is parallelled by a reduction in pTOM13 mRNA. In these situations the level of ethylene was measured. In antisense fruit the level of ethylene was reduced by approximately 80% both after wounding and during fruit ripening. Further work is in progress to define the extent of ethylene reduction as well as the effect on plant development and fruit ripening.

III. Technical Question Unanswered

It is clear from the above description that antisense RNA technology can be used successfully to control the expression of endogenous plant genes. In most instances studied, reductions of up to 95% of the expression of the target gene have been observed. Analysing several independent transformants generated from one antisense construct will make it possible to identify those plants in which the greatest inhibition has occurred. The inhibition observed in the primary transformant can then be further increased using selfing (e.g., doubling of copy number of the antisense gene). From data available in the literature and unpublished experiments in our laboratory, it is not yet clear whether complete inhibition of expression of a target gene can be achieved. It is obvious that the ultimate level of reduction is most likely to be dependent on the mechanism by which antisense RNA works. It has been proposed that inhibition of gene expression is mediated through RNA–RNA interactions and intermediates. Although this proposal has been in the literature for some time, no conclusive experiments have been carried out which clearly demonstrate this.

It has been proposed that RNA–RNA interactions lead to the inhibition of gene expression through inhibition of translation, maybe through interference with the formation of the translational initiation complex, or the progression of the ribosome along the mRNA. On the other hand, antisense RNA could also interfere with transcription, RNA processing, intron excision, and RNA transport to the cytoplasm. In this context it is worth repeating our observation that maximal levels of antisense RNA inhibition of the PG gene in plants which contain only one copy of the antisense gene is seen after the onset of PG gene expression. This might indicate that initiation of transcription is not inhibited, but that transcription is inhibited only after the onset of activation of the PG gene. It will therefore be of importance to measure the rate of transcription of the target gene as well as the antisense gene over a period of time in which inhibition of target gene expression has been measured. This study will have to be extended to include the determination of the steady state levels of both sense and antisense RNA in the nuclear and cytoplasmic RNA fractions.

It will also be of importance to identify intermediates in the formation of the RNA–RNA interactions and intermediates formed during their degradation. So far, it has not been possible to identify these intermediates.

The interpretation of these observations has been that the RNA–RNA hybrids are removed extremely rapidly and extremely efficiently.

It is not yet clear whether there is a preference for sequences which can be used in antisense vector constructions and which will lead to efficient levels of inhibition. Several examples described above point to some preference for 3′ regions of a gene. It is not clear whether this can be generalised. A systematic analysis should now be carried out which will clarify this question. It will also be of importance to determine whether genomic sequences including introns can be used to construct effective antisense vectors. We have generated a series of vectors in which different portions of the PG cDNA and the PG gene have been incorporated into antisense vectors. These experiments are now in progress. It is likely, that over the next year or two the detailed mechanisms of antisense RNA inhibition of endogenous plant genes will be elucidated.

IV. Outlook

It is clear from the above description that antisense RNA technology is a very powerful tool for the generation of novel plant mutants. It should be feasible in most cases to generate plants in which the target gene has been inhibited by at least 90%. Depending on the choice of target gene, the identification of mutants can be easy (by screening the plant phenotype), or it will depend on some biochemical measurement. The novel mutant will then be useful in studying the in vivo function of the target gene. It is however possible, that the level of reduction achieved, even after doubling of the antisense gene, may not be sufficient, to determine clearly a novel plant phenotype which is easily scored. It will therefore be of great importance to stress the biochemical analysis of these novel mutants.

In most cases studied so far, the biochemical function of the target gene was known. The experiments described with pTOM13 (Hamilton et al., pers. comm.) outline a novel approach. In this case only circumstantial evidence was available for the involvement of pTOM13 in ethylene metabolism. The generation of pTOM13 antisense mutants has further supported the view that pTOM13 is involved in ethylene metabolism. These plants will now permit the biochemical determination of the function of the gene encoding pTOM13.

This approach is a powerful novel addition to the repertoire of methods available for the identification of cDNA clones available in many cDNA libraries. Already several experiments are under way in our laboratory to generate tomatoes in which cDNA clones isolated from our cDNA libraries have been used in the construction of antisense vectors. Only the future will tell whether our present optimism over the use and power of reverse genetics will be realised.

Acknowledgements

I am grateful to Andrew Hamilton and Don Grierson for making data available before publication on the pTOM13 antisense RNA experiments.

V. References

Beachy RN, Stark DM, Deom CM, Oliver MJ, Fraley RT (1987) Expression of sequences of tobacco mosaic virus in transgenic plants and their role in disease resistance. Basic Life Sci 41: 169–180

Bird CR, Smith CJS, Ray JA, Moureau P, Bevan MW, Bird AS, Hughes S, Morris PC, Grierson D, Schuch W (1988) The tomato polygalacturonase gene and ripening-specific expression in transgenic tomato plants. Plant Mol Biol 11: 651–662

Brady C, MacAlpine G, McGlasson WB, Ueda Y (1982) Polygalacturonase in tomato fruit and the induction of ripening. Aust J Plant Physiol 9: 171–178

Cornelissen M, Vandewiele M (1989) Both RNA level and translation efficiency are reduced by antisense RNA in transgenic tobacco. Nucleic Acids Res 17: 833–843

Crookes PR, Grierson D (1983) Ultrastructure of tomato fruit ripening and the role of polygalacturonase isoenzymes in cell wall degradation. Plant Physiol 72: 1088–1093

Delauney AJ, Tabaeizadeh Z, Verma DPS (1988) A stable bifunctional antisense transcript inhibiting gene expression in transgenic plants. Proc Natl Acad Sci USA 85: 4300–4304

DellaPenna D, Alexander DC, Bennet AB (1986) Molecular cloning of tomato fruit polygalacturonase: analysis of polygalacturonase mRNA levels during ripening. Proc Natl Acad Sci USA 83: 6420–6424

Ecker JR, Davis RW (1986) Inhibition of gene expression in plant cells by expression of antisense RNA. Proc Natl Acad Sci USA 83: 5372–5376

Finkelstein R, Estelle M, Martinez-Zapater J, Somerville C (1988) *Arabidopsis* as a tool for the identification of genes involved in plant development. In: Verma DPS, Goldberg RB (eds) Temporal and spatial regulation of plant genes. Springer, Wien New York, pp 1–25 [Dennis ES et al (eds) Plant gene research. Basic knowledge and application]

Green PJ, Pines O, Inouye M (1986) The role of antisense RNA in gene regulation. Annu Rev Biochem 55: 569–597

Grierson D, Tucker GA (1983) Timing of ethylene and polygalacturonase synthesis in relation to the control of tomato ripening. Planta 157: 174–179

Grierson D, Tucker GA, Keen J, Bird CR, Schuch W (1986) Sequencing and identification of a cDNA clone for tomato polygalacturonase. Nucleic Acids Res 14: 8595–8603

Harland R, Weintraub H (1985) Translation of mRNA injected into *Xenopus* oocytes is specifically inhibited by antisense RNA. J Cell Biol 101: 1094–1099

Hastie ND, Held WA (1978) Analysis of mRNA populations by cDNA–mRNA hybrid-mediated inhibition of cell-free protein synthesis. Proc Natl Acad Sci USA 75: 1217–1221

Hemenway C, Fang R-X, Kaniewski WK, Chua N-H, Tumer NE (1988) Analysis of the mechanism of protection in transgenic plants-expressing the potato virus X coat protein or its antisense RNA. EMBO J 7: 1273–1280

Herskowitz I (1987) Functional inactivation of genes by dominant negative mutations. Nature 329: 219–222

Hobson GE (1964) Polygalacturonase in normal and abnormal tomato fruit. Biochem J 92: 324–332

Hobson GE (1965) The firmness of tomato fruit in relation to polygalacturonase activity. J Horticult Sci 40: 66–72

Holdsworth MJ, Bird CR, Ray J, Schuch W, Grierson D (1987) Structure and expression of an ethylene related mRNA from tomato. Nucleic Acids Res 15: 731–739

Holdsworth MJ, Schuch W, Grierson D (1988) Organisation and expression of a wound/ripening related small multigene family from tomato. Plant Mol Biol 11: 81–88

Izant JG, Weintraub H (1984) Inhibition of thymidine kinase gene expression by antisense RNA: a molecular approach to genetic analyses. Cell 36: 1007–1015

Loesch-Fries LS, Halk E, Merlo D, Jarvis N, Nelson S, Krahn K, Burhop L, Arntzen CJ, Ryan C (1987) Molecular strategies for crop protection, expression of alfalfa mosaic virus coat protein gene and antisense cDNA in transformed tobacco tissue. UCLA Symp Proc Mol Cell Biol NS 48: 221–234

Kuhlemeier C, Green PJ, Chua NH (1987) Regulation of gene expression in higher plants. Annu Rev Plant Physiol 38: 221–57

Paterson BM, Roberts BE, Kuff EL (1977) Structural gene identification and mapping by DNA–mRNA hybrid-arrest cell-free translation. Proc Natl Acad Sci USA 74: 4370–4374

Pines O, Inouye M (1986) Antisense RNA regulation in prokaryotes. Trends Genet 2: 284–287

Ray J, Knapp JE, Grierson D, Bird C, Schuch W (1988) Identification and sequence determination of a cDNA clone for tomato pectin esterase. Eur J Biochem 174: 119–124

Rezaian AM, Skene KGM, Ellis JG (1988) Antisense RNAs of cucumber mosaic virus in transgenic plants assessed for control of the virus. Plant Mol Biol 11: 463–471

Robert LS, Donaldson PA, Ladaique C, Altosaar I, Arnison PG, Fabijanski SF (1989) Antisense RNA inhibition of β-glucuronidase gene expression in transgenic tobacco plants. Plant Mol Biol 13: 399–409

Rodermel SR, Abbott MS, Bogorad L (1988) Nuclear–organelle interactions: nuclear antisense gene inhibits ribulose bisphosphate carboxylase enzyme levels in transformed tobacco plants. Cell 55: 673–681

Rothstein SJ, DiMaio J, Strand M, Rice D (1987) Stable and heritable inhibition of the expression of nopaline synthase in tobacco expressing antisense RNA. Proc Natl Acad Sci USA 84: 8439–8443

Sandler SJ, Stayton M, Townsend JA, Ralston ML, Bedbrook JR, Dunsmuir P (1988) Inhibition of gene expression in transformed plants by antisense RNA. Plant Mol Biol 11: 301–310

Schuch W, Bird CR, Ray J, Smith CJS, Watson CF, Morris PC, Gray JE, Arnold C, Seymour GB, Tucker GA, Grierson D (1989) Control and manipulation of gene expression during tomato fruit ripening. Plant Mol Biol 13: 303–311

Seymour GB, Harding SE (1987) Molecular size of tomato (*Lycopersicon esculentum* Mill) fruit polyuronides by gel filtration and low-speed sedimentation equilibrium. Biochem J 245: 463–466

Sheehy RE, Pearson J, Brady CJ, Hiatt WR (1987) Molecular characterisation of tomato fruit polygalacturonase. Mol Gen Genet 208: 30–36

Sheehy RE, Kramer MK, Hiatt WR (1988) Reduction of polygalacturonase activity in tomato fruit by antisense RNA. Proc Natl Acad Sci USA 85: 8805–8809

Simons RW, Kleckner N (1988) Biological regulation by antisense RNA in prokaryotes. Annu Rev Genet 22: 567–600

Slater A, Maunders MJ, Edwards K, Schuch W, Grierson D (1985) Isolation and characterisation of cDNA clones for tomato proteins. Plant Mol Biol 5: 137–147

Smith CJS, Slater A, Grierson D (1986) Rapid appearance of an mRNA correlated with ethylene synthesis encoding a protein of molecular weight 35,000. Planta 168: 94–100

Smith CJS, Watson CF, Ray J, Bird CR, Morris PC, Schuch W, Grierson D (1988) Antisense RNA inhibition of polygalacturonase gene expression in tomatoes. Nature 334: 724–726

Smith CJS, Watson CF, Morris PC, Bird CR, Seymour GB, Gray JE, Arnold C, Tucker GA, Schuch W, Grierson D (1990) Inheritance and effect on ripening of antisense polygalacturonase genes in transgenic tomatoes. Plant Mol Biol (in press)

Smithies O, Gregg RG, Boggs SS, Koralewski MA, Kucherlapati RS (1985) Insertion of DNA sequences into the human chromosomal β-globin locus by homologous recombination. Nature 317: 230–234

Thomas H, Grierson D (1987) Developmental mutants in higher plants. Cambridge University Press, Cambridge

Tucker GA, Grierson D (1982) Synthesis of polygalacturonase during tomato fruit ripening. Planta 155: 64–67

van der Krol AR, Lenting PE, Veenstra J, van der Meer IM, Koes RE, Gerats AGM, Mol JNM, Stuitje AR (1988a) An antisense chalcone synthase gene in transgenic plants inhibits flower pigmentation. Nature 333: 866–869

van der Krol AR, Mol JNM, Stuitje AR (1988b) Modulation of eukaryotic gene expression by complementary RNA or DNA sequences. BioTechniques 6: 958–976

Weising K, Schell J, Kahl G (1988) Foreign genes in plants: transfer, structure, expression, and applications. Annu Rev Genet 22: 421–477

Chapter 7

Molecular Biology of Flavonoid Pigment Biosynthesis in Flowers

Trevor W. Stevenson

Calgene Pacific Pty Ltd, Collingwood, Victoria 3066, Australia

With 4 Figures

Contents

I. Introduction

For some time considerable scientific and commercial interest has been generated by the potential impact of genetic engineering technology on the broad acre agricultural crops. It is only recently that the application of this technology to the development of new varieties of flowers has been seriously considered. This is somewhat surprising given the size of the cut flower industry, the small number of plant species which dominate the cut flower market and the fact that a considerable body of genetic and physiological information is available describing economically important traits such as colour, vase life and disease response.

The world cut flower industry is centered on Holland, with over 90% of the intra-European trade, and 60% of world export trade in cut flowers handled by the auction centers in Aalsmeer and Westlands. According to data published by the International Association of Horticultural Producers, the total cut flower sales by Dutch auction houses was in excess of US$ 3 billion in 1989, and is increasing at a rate of approximately 5% per year. The sale of cut flowers is dominated by three species: rose, chrysanthemum, and carnation, which together account for nearly 50% of all flowers sold. It is not surprising therefore that these species have been subject to considerable selection and cultivar development.

Some ornamentals, such as bedding plants like petunia, are produced by seed propagation. However, much of the production of high value cut flower varieties is by vegetative or clonal propagation. This type of propagation is well adapted to the use of breeding and selection schemes where a large number of progeny are easily screened to find the individual plant or plants which display those traits required by the market. Individual plants which display the desired phenotype are multiplied by vegetative propagation for subsequent flower production. Compared to propagation by seed, clonal or vegetative propagation is expensive and time consuming, however, the high value of a number of ornamental species makes these breeding and production techniques commercially viable.

Even to the casual observer the efforts of these traditional breeding and selection programs have clearly been successful, with many of the spectacular "products" dominating flower sales and gardens everywhere. There are, however, several quite severe limitations to such breeding and selection programs and the most obvious constraint is the natural barriers preventing different species being hybridized. Essentially this limits the available gene pool and as a consequence limits the range of phenotypes able to be developed. The application of recombinant DNA technology and gene transfer methods have the potential to overcome these incompatibility barriers.

Another problem with conventional plant breeding is that without extensive and time consuming back-crossing it is difficult to alter one specific trait only. Direct and specific gene transfer methods have the potential to overcome this limitation. Elite, commercial cultivars can be altered for one trait only. For example, new flower colours can be produced from an existing elite, commercially successful cultivar without altering any of the other characteristics of that plant. Thus the application of recombinant DNA technology provides an opportunity to alter ornamental plants in a highly directed manner.

The application of recombinant DNA technology is always dependent upon:

(*i*) an understanding of the target biochemical pathway or process to be altered,
(*ii*) the ability to isolate the gene or the genes responsible for the biochemical step, or steps, to be manipulated, and
(*iii*) the ability to transfer and express the genes of interest in the plant species to be genetically altered.

Whilst there have been intensive breeding programs developed for the important flower species and much progress made in the development of new varieties, only minimal attention has been given to quantitative genetic analysis of important traits such as flower colour. Fortunately, *Petunia hybrida* and *Antirrhinum majus* have been used as model systems for genetic studies of pigment production (Wiering and de Vlaming, 1984) and the spatial distribution of flower pigments (Coen et al., 1988), and these studies form the foundation for the application of recombinant DNA strategies to the cut flower industry.

Flavonoid pigments make the greatest contribution to flower colour. In this chapter the pathway for flavonoid pigment biosynthesis is briefly outlined and recent progress towards the isolation and transfer of genes responsible for flower pigment biosynthesis is reviewed.

II. Flavonoid Pigments

The flavonoid pigments comprise a widespread group of water-soluble phenolics generally found in plant vacuoles as glycosides. They are often brightly coloured and can be red, crimson, blue, purple or yellow. The

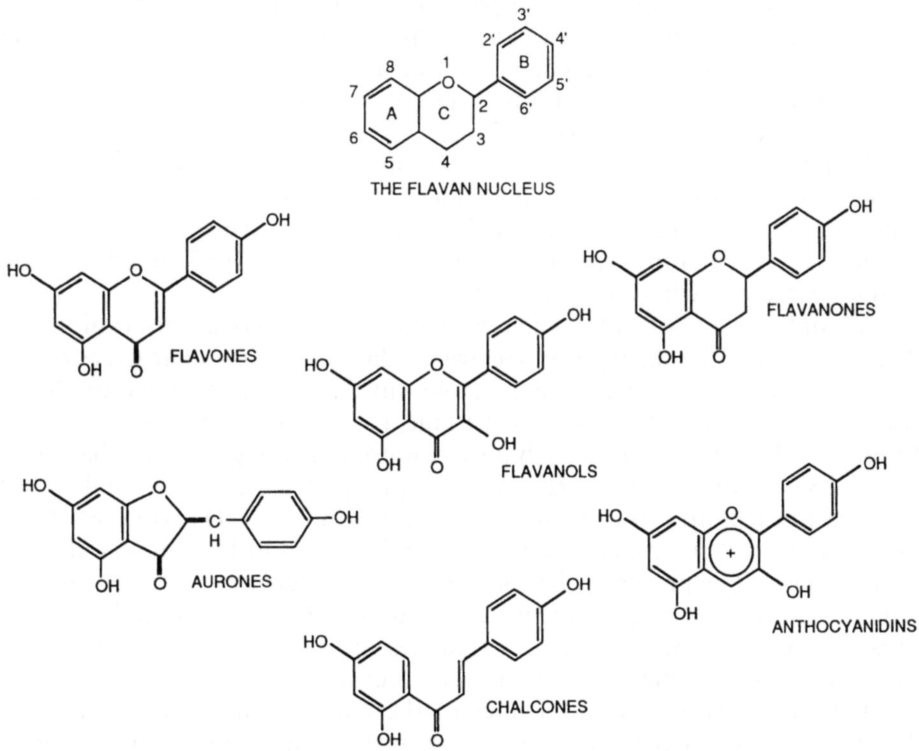

Fig. 1. Structure of various flavonoids involved in flower pigmentation

flavonoid structure is based on two aromatic rings joined by a three carbon unit. The varying degree of oxidation of this central three carbon unit (C-ring) gives rise to the different classes of flavonoids (Fig. 1).

Flavonols and flavones, whilst they are widespread in flower petals, do not contribute significantly to colour. However, some flavonols, when methylated or hydroxylated, do contribute a yellow colour in flowers. Examples are, 8-hydroxyflavonol, the yellow colour of the meadow pea (*Lathyrus pratensis*), gossypetin, the compound responsible for the distinctive yellow colour of cotton flowers, and quercetagetin, the 6-hydroxy derivative of quercetin, which is the principal yellow pigment in the flowers of many *Primula* and *Rhododendron* sp. (Harbourne, 1967). Whilst flavones do not contribute directly to flower colour they do act as co-pigments, intensifying the yellow of other flavonoids such as chalcones and aurones. The colourless flavones and flavonols appear to impart a white, cream or ivory colour to flowers.

Chalcones and aurones can both make major contributions to the yellow colour of petals. The presence of these pigments can easily be established by exposure to ammonia vapour, upon which the pigment turns from yellow to red. The occurrence of these pigments is restricted to about nine plant families and one example is the chalcone isosalipurposide, the predominant pigment in yellow carnations, *Dianthus caryophyllus* (Spribille and Forkmann, 1982). The chalcone and aurone pigments, however, are often associated with carotenoids in petals and because of this may only partly contribute to the yellow colour.

Anthocyanidins are the class of flavonoid which makes the greatest contribution to flower colour. The names of the most common anthocyanidins, cyanidin, pelargonidin and delphinidin are derived from the flower species from which they were first isolated (Harbourne, 1967) and differ only in the degree of hydroxylation of the B-ring (Fig. 2).

A number of factors affect the relative contribution of anthocyanidins to flower colour. The absorption spectra of flavonoids are characterized by two separate bands, one at longer visible wavelengths determined by the B-ring and the second, the ultra violet range, is determined by conjugation of the A-ring. In acid solution the highly coloured anthocyanin salts have one main absorption maxima in the visible region, occurring between 465 and 550 nm. For example, the absorption maxima (in 0.01% HCl) of scarlet-red pelargonidin is 520 nm, crimson cyanidin 535 nm and blue-mauve delphinidin 546 nm. The presence of additional hydroxyl groups on the B-ring has a major effect on colour. With the addition of OH groups to the B-ring the presence of non bonding electrons causes an increase in the absorption maximum, shifting the colour from red to crimson towards blue.

Anthocyanins are the glycosides of the anthocyanidin chromophores. The presence of sugar moieties attached to the 3-carbon position of the C-ring confers upon anthocyanins a high degree of solubility and stability.

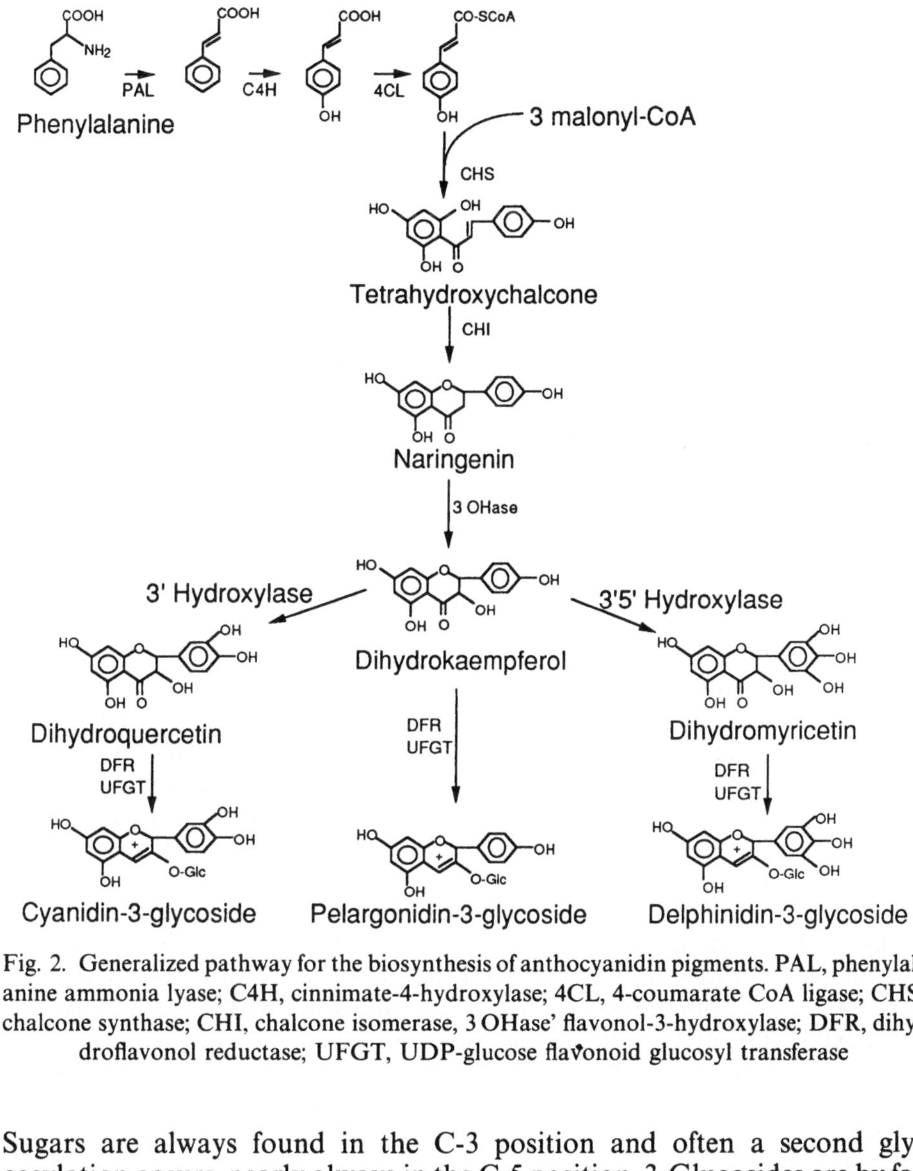

Fig. 2. Generalized pathway for the biosynthesis of anthocyanidin pigments. PAL, phenylalanine ammonia lyase; C4H, cinnimate-4-hydroxylase; 4CL, 4-coumarate CoA ligase; CHS, chalcone synthase; CHI, chalcone isomerase, 3 OHase' flavonol-3-hydroxylase; DFR, dihydroflavonol reductase; UFGT, UDP-glucose flavonoid glucosyl transferase

Sugars are always found in the C-3 position and often a second glycosylation occurs, nearly always in the C-5 position. 3-Glucosides are by far the most common of all anthocyanins with cyanidin 3-glucoside occurring in many flowers, leaves and fruits. The 3-glucosides of pelargonidin, cyanidin, and delphinidin can all be readily isolated from varieties of Chinese primrose, *Primula sinensis* (Harbourne, 1967). 3-Galactosides are much less common than the 3-glucosides, but do occur in the berries of *Vaccinium* species (e.g., cranberry, blueberry). Glycosylation of anthocyanidins has a hypochromic effect on the spectra with slight increases in absorption maxima occurring. Compared with other modifications of the flavan nucleus (Fig. 1), the nature of the sugar residue has only a minor influence on pigment colour.

The colour of anthocyanins is very sensitive to pH, being orange and red in strongly acidic media and reddish-violet to violet in weakly acidic or neutral solutions. A blue colour can only be produced in an alkaline environment. As plant vacuoles, the intra-cellular location of anthocyanin pigments, are invariably acid (Matile, 1975), anthocyanins cannot produce stable blue colour in vivo unless there is a mechanism for stabilization or colour modification. One mechanism for stabilization and colour modification is self association. Asen et al. (1972) found that the absorbance of cyanidin 3-5-diglucoside increased up to 300 times when the concentration was increased from 10^{-4} M to 10^{-2} M (100 times) and suggested that self association of the anthocyanins was the cause.

The amount of anthocyanin present in the petal can have profound effects on the intensity of pigment colour. For example, low pigment content (0.01% dry weight) results in very faint pinkish petals in the rose variety "Madame Butterfly", whereas very high concentrations (up to 15% dry weight) result in deep purple-black colours such as in the tulip variety "Queen of the Night" or the pansy variety "Jet Black" (Harbourne, 1967). Methylation of the B-ring of anthocyanin pigments can occur and the methylated derivatives, peonidin, petunidin and malvidin are widespread in flowers. It is quite rare for the 4′ position of the B-ring to be methylated. Methylation of the anthocyanidins causes a slight shift in the absorption maxima towards red colours. However, the effects of methylation are usually obscured by other factors such as vacuolar pH and copigmentation.

Shifts in the absorption spectrum of anthocyanins in the presence of flavones and tannins were first observed in 1931 and the phenomena was described as co-pigmentation (Robinson and Robinson, 1931). The effects of co-pigmentation depend on the type and concentration of anthocyanin, the type and concentration of flavone and the pH of the anthocyanin environment. The co-pigmentation effects of complexing with flavones, or by complexing with metals such as magnesium, iron or aluminium, can have quite marked effects on the final colour of flavonoid pigments. In some cases, such as blue cornflower *Centaurea cyanus*, the pigment is cyanin complexed with polysaccharide as well as ferric and aluminium ions (Goto, 1987). The blue colour in this case is due to non-delphinidin type anthocyanidin derivatives.

However, blueness in plants is nearly always due to delphinidins occurring in association with metals or other flavonoids. Blue flowers are, however, rare and are difficult to introduce into a plant variety unless the "parent" or wild type plant also has blue flowers, hence the absence of a "true-blue" rose or snapdragon. The blue-toned, purple rose varieties presently available, as with the blue cornflower *C. cyanus*, contain cyanidin, probably co-pigmented with leuco-anthocyanidin. The difficulty with breeding a pure blue rose lies in the fact that no source of delphinidin exists in the genus *Rosa*. Cyanidins are universally present and delphinidin is absent and cannot be obtained by normal genetic means.

III. Biosynthesis of Flavonoids

The main precursor of all flavonoid compounds is phenylalanine. This aromatic amino acid forms the basis of the B-ring in the final flavonoid structure, whereas the A-ring originates from acetate via malonyl CoA. The early steps in flavonoid biosynthesis are part of the phenylpropanoid pathway and are common to the biosynthesis of lignins, cinnamic esters, coumarins, phytoalexins as well as flavonoids. The first step, deamination of phenylalanine, is catalysed by the enzyme phenylalanine ammonia lyase (PAL), the immediate product being *trans*-cinnamic acid (see Fig. 2). The second step involves the hydroxylation of *trans*-cinnamic acid at C-4, and this reaction is catalyzed by cinnimate 4-hydroxylase (C4H), a cytochrome P-450 mixed function mono-oxygenase. The product of this reaction is *p*-coumaric acid. The next enzymatic step is carried out by 4-coumarate: CoA ligase (4CL) and has been extensively studied in petunia (Ranjeva et al., 1976) and is the last step in the phenylpropanoid pathway.

The first reaction specific to the biosynthesis of flavonoids involves the condensation 4-coumaroyl-CoA with three molecules of malonyl-CoA. The immediate product of this reaction was not conclusively identified as a chalcone until 1980 (Heller and Hahlbrock, 1980) at which time the enzyme responsible was renamed chalcone synthase (CHS). Normally the chalcone product of this reaction, 4,2′,4′,6′ tetrahydroxy chalcone, is rapidly isomerized to the flavanone, naringenin. This isomerization is catalysed by the enzyme chalcone isomerase (CHI), but can also occur spontaneously. Naringenin is the common precursor for a number of flavonoid derivatives, including anthocyanidins. Hydroxylation of naringenin at the 3 position of the C-ring to produce dihydrokaempherol is performed by a soluble enzyme, flavonol 3-hydroxylase (3-OHase). With the exception of 4,2′,4′,6′ tetrahydroxy chalcone, which when present in the glycosylated form is yellow, all the pigment precursors mentioned so far are colourless.

The degree of hydroxylation of the B-ring of anthocyanidins has a significant effect on the colour of the final pigments. Further hydroxylation of the B-ring does not always occur and the 4′ hydroxylated dihydrokaempherol can be reduced and glycosylated to produce pelargonidin pigments. However, the B-ring of dihydrokaempherol can be further hydroxylated at the 3′ position, producing dihydroquercetin and ultimately cyanidin pigments. Hydroxylation may also occur at both the 3′ and 5′ positions converting either the 4′ hydroxylated dihydrokaempherol or the 3′4′

hydroxylated dihydroquercetin to the 3′4′5′ hydroxylated dihydromyricetin, the precursor to delphinidin pigments.

In petunia there exist two distinct enzymes which carry out these hydroxylation reactions, a 3′ hydroxylase and a 3′5′ hydroxylase. These enzymes are encoded by distinct and unlinked loci, and their expression has a marked effect on the type of pigments which accumulate in the petal. Stotz et al. (1985) found a strict correlation between the presence of the Ht1 gene and the presence of a 3′ hydroxylase enzyme capable of hydroxylation of either naringenin or dihydrokaempherol. The 3′ hydroxylase enzyme appeared to be associated with the microsomal fraction and required NADPH as a cofactor. The expression of the Hf1 gene in petunia can result in almost all of the flavonoid products being hydroxylated at both the 3′ and 5′ position on the B-ring, with the predominant pigments being delphinidin derivatives (Gerats et al., 1982). Characterization of the 3′5′ hydroxylase enzyme in *Verbena hybrida* flowers (Stotz and Forkmann, 1982), revealed that this enzyme was also associated with the microsomal membrane fraction and required NADPH as a cofactor.

Regardless of the pattern of hydroxylation of the B-ring, the dihydroflavonols (dihydroquercetin, dihydrokaempherol and dihydromyricetin) are all colourless, and accumulation of these pigments in flowers results in white or cream colours (Gerats et al., 1982). Conversion of dihydroflavonols into coloured anthocyanins requires at least two further enzymes: dihydroflavonol reductase (DFR), which carries out a reduction of the C ring and a UDP-glucose flavonoid glucosyl-transferase (UFGT). As mentioned previously, anthocyanidins themselves are unstable and are rarely present in high amounts, however, glycosylation of the 3-hydroxy group of the C ring gives these compounds solubility as well as chemical stability. The last steps in the biosynthesis of anthocyanidins are essentially further substitutions of the basic molecule by glycosylation, rhamnosylation, acylation, and methylation, all of which make relatively minor contributions to the final flower colour, changing the intensity of the pigments rather than affecting the predominant colour of the pigments.

A. Flavonoid Biosynthesis Genes

A number of genes coding for flavonoid biosynthesis enzymes have been cloned (Mol et al., 1988) and there continues to be considerable effort directed towards isolating the other genes of the pathway. Genes have been isolated from a number of sources, the most common being parsley (*Petroselinum hortense*), bean (*Phaseolus vulgaris*), snapdragon (*Antirrhinum majus*), maize (*Zea mays*), and petunia (*Petunia hybrida*). Table 1 lists the plant flavonoid biosynthesis genes cloned to date.

Table 1. Summary of flavonoid biosynthesis genes isolated

Phenylalanine ammonia lyase	*Petroselium hortense*	Kuhn et al., 1984
	Phaseolus vulgaris	Edwards et al., 1985
4 Coumarate CoA ligase	*Petroselium crispum*	Douglas et al., 1987
	P. hortense	Kreuzaler et al., 1983
Chalcone synthase	*P. crispum*	Herrmann et al., 1988
	Antirrhinum majus	Sommer and Saedler, 1986
	Zea mays	Wienard et al., 1986
	Petunia hybrida	Koes et al., 1986
	Phaseolus vulgaris	Ryder et al., 1987
	Ranunculus acer	Niesback-Klosgen et al., 1987
	Hordeum vulgare	Niesback-Klosgen et al., 1987
	Magnolia liliflora	Niesback-Klosgen et al., 1987
	Arabidopsis thaliana	Feinbaum and Asubel, 1988
Chalcone isomerase	*Phaseolus vulgaris*	Mehdy and Lamb, 1987
	Petunia hybrida	van Tunen et al., 1988
Dihydroflavonol reductase	*Zea mays*	O'Reilly et al., 1985
	Antirrhinum majus	Martin et al., 1985
	Petunia hybrida	Beld et al., 1989
UDP glucose-flavonoid glucosyl transferase	*Zea mays*	Fedoroff et al., 1984
Bz2	*Z. mays*	McLaughlin and Walbot, 1987
C1 regulatory protein	*Z. mays*	Paz-Ares et al., 1986
Lc regulatory protein	*Z. mays*	Ludwig et al., 1989
B regulatory protein	*Z. mays*	Chandler et al., 1989

The cloning of genes involved in flavonoid biosynthesis has been facilitated by two features of flavonoids. Firstly, biosynthesis can be induced to relatively high levels in isolated cell cultures, therefore providing a pre-induction for sequences or gene clones. Secondly, there exist several well developed genetic systems where a number of structural gene and regulatory gene mutations are available. The best characterized systems are those in *Antirrhinum* (Coen et al., 1988), maize (Dooner, 1983), and petunia (Weiring, 1974).

The first flavonoid biosynthesis gene isolated was chalcone synthase (CHS). Using UV induction of dark-grown parsley suspension cells, Kreuzaler et al. (1983) isolated and identified the CHS cDNA by differential hybridization, hybrid arrest and hybrid select translation. A different approach was used to isolate a clone for the CHI gene (Mehdy and Lamb, 1987; van Tunen et al., 1988) where antibodies raised against purified CHI protein were used to screen cDNA libraries cloned in expression vectors. Using similar approaches cDNA probes for PAL (Kuhn et al., 1984) have also been isolated. With the availability of these initial cDNA clones, heterologous clone isolation has been possible and clones for CHS, and PAL have been isolated from a number of different species (Table 1).

To isolate genes by the differential expression strategy it is essential that the target gene be expressed at a reasonable level. CHS is the most highly expressed flavonoid biosynthesis gene and at maximal induction comprises almost 2% of the total rate of protein synthesis (Schroder et al., 1979) and this high abundance undoubtedly contributed to its "ease" of isolation. However, the differential screening approach to gene isolation is not feasible when attempting to isolate low abundance structural or regulatory genes. Fortunately, the well characterized genetics of *Antirrhinum* and maize, as well as the presence in these species of mobile genetic elements, has made possible the isolation of a number of other flavonoid biosynthesis genes.

Three mobile genetic elements have been characterized in *Antirrhinum*, Taml (Bonas et al., 1984), Tam2 (Upadhyaya et al., 1985), and Tam3 (Sommer et al., 1985). These elements have been shown to affect several loci involved in flavonoid pigment biosynthesis, in paticular *nivea* and *pallida*. The *nivea* locus encodes the CHS gene (Strickland and Harrison, 1974) and molecular analysis of various mutant alleles has detailed the mechanisms by which these mobile elements affect CHS gene activity (Coen et al., 1986). The mobile element Tam3 has also been used to isolate the gene encoding the *pallida* locus, (Martin et al., 1985). Subsequent sequence analysis showed strong homology of the *pallida* clone to the A1 or DFR gene in maize (Coen et al., 1986). The petunia DFR gene has since been isolated using the *pallida* genomic DNA fragment isolated from *Antirrhinum* as a heterologous probe (Beld et al., 1989) and by differential screening (Stevenson et al., 1990 unpubl. data).

Two mobile element systems in maize, En-Spm (Pereira et al., 1985, 1986) and Ac-Ds (Fedoroff, 1983), have been used to clone genes involved in anthocyanin production in the aleurone layer of seed. Production of anthocyanin pigments in maize requires the action of a number of genes, both regulatory and structural. Of the structural genes, the A1 locus has been cloned and shown to encode dihydroquercetin reductase (O'Reilly et al., 1985; Schwarz-Sommer et al., 1987), C2 appears to encode the structural gene for CHS (Wienand et al., 1986) and Bz1 encodes UDP glucose: flavonoid 3-0-glucosyl transferase (UFGT) (Fedoroff et al., 1984; Dooner and Nelson, 1977). A number of other loci involved in pigment biosynthesis are affected by transposons (Nevers et al., 1986), and are therefore candidates to be cloned by transposon tagging. The Bronze2 locus (Bz2) still has no identified gene product, but the gene has been cloned (McLaughlin and Walbot, 1987). The first regulatory gene isolated was the

C1 gene (Paz-Ares et al., 1986) which was found to have homology to the protein products of mammalian oncogenes (Paz-Ares et al., 1987). Recently the gene Lc, a member of the R regulatory gene family, has been cloned and shown to have homology to *myb* c, a putative transcription factor (Ludwig et al., 1989). The B regulatory gene has since been isolated utilizing sequence homology to the R gene (Chandler et al., 1989).

The number of cloned probes available for flavonoid pigment biosynthesis genes is rapidly expanding. An interesting feature of the clones presently available is the absence of genes coding for two of the key enzymes involved in determining the final colour of the anthocyanin pigments, the 3′ hydroxylase and the 3′5′ hydroxylase. As distinct from all of the other cloned genes, these two hydroxylases, and the still uncloned cinnimate 4-hydroxylase, are microsomal membrane proteins apparently sharing many characteristics of cytochrome P-450 enzymes (Russell, 1971; Stotz et al., 1985; Stotz and Forkman, 1982; Hagmann et al., 1983). Estimates from our laboratory, and from other workers (Larson and Bussard, 1986; Hagmann et al., 1983), suggests that the total cytochrome P-450 content of petals and maize seedlings is about 0.20 nmol/mg protein, indicating that these enzymes are present at quite low levels, even in tissues where anthocyanidin biosynthesis is high. The low level of total cytochrome P-450 enzymes, and presumably significantly lower levels of individual substrate-specific, cytochrome P-450 enzymes, may explain why these genes have yet to be cloned using differential screening techniques. The inherent instability and lability characteristic of plant cytochrome P-450s is evident from the small number of plant cytochrome P450 enzymes which have been characterized and/or purified. Purification of a cytochrome P-450 has been reported from tulip, *Tulipa gesneriana* (Higashi et al., 1985), Jerusalem artichoke, *Helianthus tuberosis* (Gabriac et al., 1985), and Avocado, *Persea americana* (O'Keefe and Leto, 1989). The low abundance and labile nature of cytochrome P-450 proteins makes gene cloning via antibody screening or protein sequence and oligonucleotide screening difficult. Gene isolation via transposon tagging in the maize or *Antirrhinum* systems are, however, potential means of isolating these genes. Both maize and *Antirrhinum*, however, only synthesize pelargonidin and cyanidin pigments and not delphinidins, and as such apparently lack the 3′5′ hydroxylase gene. Several groups, including our own, are actively trying to isolate the 3′ hydroxylase and 3′5′ hydroxylase genes which play such an important role in determining the final colour of flowers.

B. Genetic Engineering of Flavonoid Biosynthesis

The availability of cloned copies of a number of flavonoid pigment biosynthesis genes has made feasible the application of gene manipulation technoiogy to the production of new and novel flowers. However, the ability to routinely transfer and express genes of interest into a particular target crop is still one of the major hurdles to the application of this technology. The availability of a well established transformation system (Horsch et al., 1985), and well characterized mutant lines, many of which are defined with respect to flavonoid pigments, made petunia the plant species in which the initial flavonoid gene manipulation experiments have been performed.

There are essentially two different ways to manipulate flower colour using cloned pigment biosynthesis genes:

(*i*) Through the expression of pigment biosynthesis genes from a heterologous species in a target plant which does not normally express that particular gene, the result being the production of new pigments not normally synthesized in that target flower.
(*ii*) Through highly specific inhibition of the expression of a particular flavonoid gene or genes, the result being a blockage in the pigment biosynthesis pathway and the accumulation of precursors or flavonoid pigments not previously abundant.

The first successful demonstration of the molecular manipulation of flower colour was reported by workers from the Max-Planck Institute in Cologne in 1987 (Meyer et al., 1987). These workers expressed the maize DFR gene in petunia plants and created a new pigment biosynthesis pathway leading to production of pelargonidin pigments which are not normally found in petunia. The petunia enzyme responsible for the conversion of colourless dihydroflavonols to anthocyanins, DFR, is unusual in that it exhibits high substrate specificity with regard to B-ring substitution pattern (Forkmann and Ruhnau, 1987). The 3′4′5′ hydroxylated dihydromyricetin is efficiently reduced to ultimately produce delphinidin products, the 3′4′ hydroxylated dihydroquercetin is only poorly reduced to cyanidin products and the 4′ hydroxylated dihydrokaempferol is not reduced by DFR at all. These differences in substrate specificity of the petunia enzyme are entirely consistent with the end products of flavonoid biosynthesis found in petunia flowers (Gerats et al., 1982). The absence, or very rare occurrence, of pelargonidin pigments in petunia is a result of the relative substrate affinities of the petunia DFR for the various hydroxylated forms of dihydroflavonols.

Maize is able to synthesize anthocyanin pigments in the aleurone layer of the seed. One of the gene loci controlling the biosynthesis of the final anthocyanidin pigments, A1, encodes dihydroquercetin reductase (Reddy et al., 1987), the enzyme which carries out the conversion of dihydroquercetin to leucocyanidin as well as the conversion of dihydrokaempherol to leucopelargonidin, the precursor to pelargonidin pigments. Meyer et al. (1987) placed the maize A1 coding sequence behind the constitutive CaMV35S promoter and transferred this gene construct into a petunia variety which accumulated dihydrokaempherol. Expression of the A1 gene product in transgenic petunia resulted in petals which contained pelargonidin pigments and were a new and novel colour. In this first demonstration of the manipulation of flower colour by genetic engineering these workers had created in petunia a new flower pigment biosynthesis pathway and the transgenic plants accumulated flavonoid pigments not previously present in that plant species. This was the first demonstration of the introduction of a new pigment biosynthetic pathway in plants via direct gene transfer.

The second approach to the manipulation of flower pigments by genetic engineering is to block the completion of flavonoid synthesis by antisense expression of one of the genes in the biosynthetic pathway. Antisense inhibition of gene expression is based on blocking the information flow from DNA via RNA to functional protein by the expression and presence of an RNA strand complementary to at least part of the sequence of the target gene mRNA. Presumably through specific hybridization between the antisense mRNA and the sense mRNA, and via mechanisms not yet understood, the mRNA/antisense mRNA duplex is either rapidly degraded or nuclear processing of the mRNA is impaired. As a result, functional expression of the target gene is inhibited. The specific use of antisense gene expression to modify flower colour was first demonstrated by van der Krol and colleagues at The Free University in the Netherlands (van der Krol et al., 1988).

Similar work in our laboratory using antisense expression has clearly indicated the power of this technology in the production of new flower colour phenotypes (Fig. 3). By placing a full length petunia CHS cDNA fragment in an antisense orientation behind the constitutive CaMV 35S promoter and expressing this antisense construct in transgenic petunia (cv. Old Glory Blue), transgenics were obtained which were completely blocked in pigment production in the petals.

Northern blot analysis of RNA isolated from the leaves and petals of the antisense transgenic plant (Fig. 4) indicate an almost complete absence of chalcone synthase mRNA as compared to the normal, non-transgenic plant. In transgenic plants the mRNA levels for the other genes in the flavonoid biosynthesis pathway, PAL, CHI, and DFR, were normal, with steady state levels of mRNA for these genes being comparable in the normal and transgenic plant. Two of these genes, CHI and DFR, code for enzymes responsible for steps later in the pigment biosynthetic pathway. It is interesting to note that expression levels of these two genes are unaltered irrespective of the absence of precursors for that particular biochemical step.

Whilst the results of antisense CHS gene expression shown in Fig. 3 appear to provide a complete and phenotypically marked effect, considerable variation in the phenotype of CHS antisense construct containing flowers has been observed. In a detailed study of the variation which can be found among transgenic antisense CHS plants, van der Krol et al. (1990b)

Fig. 3. Petunia cv. Old Glory Blue and transgenic petunia cv. Old Glory Blue expressing the CaMV35S-antisense chalcone synthase (CHS) gene construct (Stevenson et al., 1990 unpubl. data)

observed pigmentation patterns which ranged from wild type or unaltered through to complete inhibition of pigment biosynthesis similar to that in Fig. 3. Consistent with our findings, only the CHS steady state mRNA level was affected, with expression of the other pigment biosynthesis genes being unaltered. Reduction of pigment levels appeared to correlate well with the observed decreases in CHS mRNA, as did the spatial distribution of pigment inhibition. The steady state level of antisense CHS transcript, however, appeared not to correlate with flower phenotype. Flowers exhibiting various degrees of pigmentation showed almost equivalent levels of antisense CHS mRNA, and flowers with a complete absence of pigment had undetectable levels of antisense CHS mRNA. The effects of the antisense construct were usually consistent within a given plant, however, some antisense CHS plants did show variation in pigmentation over time. Amongst agents shown to influence the pigmentation patterns of the antisense plants were the phytohormone gibberellic acid (GA-3) and lighting conditions. Spraying with GA-3 increased the number of pigmented cells, and light supplementation lead to decreased levels of pigmentation. The chromosomal localization of the antisense CHS genes indicated a bias towards chromosome I, but neither a phenotype, nor level of antisense gene

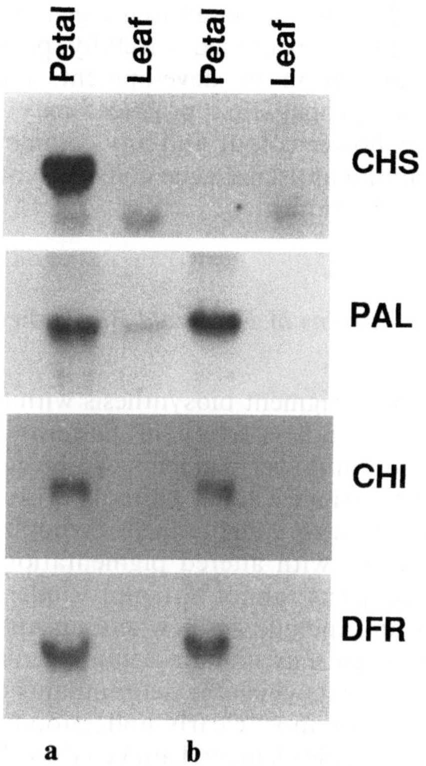

Fig. 4. Analysis of petunia cv. Old Glory Blue (a) and transgenic petunia cv. Old Glory Blue expressing the antisense CHS gene construct (b). Northern blots were loaded with 10 μg total RNA from petal and leaf. The CHS probe was a ^{32}P-UTP labelled antisense strand RNA probe such that endogenous CHS mRNA levels were determined. PAL, CHI and DFR probes were ^{32}P-labelled cDNA fragments

expression, appeared to correlate with regions of integration. The authors suggest that the different phenotypes and physiological sensitivities of various antisense CHS phenotypes is due to the DNA sequences near or flanking the site of integration of the antisense CHS gene construct.

The use of antisense expression is not confined to the CHS gene. As mentioned previously, the number of cloned pigment biosynthesis genes is growing rapidly, and each of these genes are potential candidates to be expressed in an antisense orientation. Strategic antisense expression of these other pigment biosynthesis genes could result in subtle colour modifications of existing elite varieties or create new and novel flower colours. In some cases where pigment biosynthesis genes are not available, the isolation of flavonoid biosynthesis genes from the particular species to be modified may be required. This may be possible using heterologous gene probes.

Quite clearly, the phenomenon of antisense gene-mediated alteration of flower colour needs to be examined carefully prior to embarking on commercial scale propagation or development. Factors such as light regimes or commercial propagation practices may influence phenotype characteristics such as flower colour and any commercial development of flower varieties containing antisense gene constructs will obviously require extensive trials and evaluation.

C. Patterns of Flavonoid Biosynthesis

The level and intensity of pigment biosynthesis within flowers is often not uniform, leading to an endless array of patterns or arrangements of pigments. This variation in flower pigments can be considered to be either clonal or non-clonal in origin. Clonal patterns arise due to genetic alterations which are passed on to daughter cells through mitotic cell division, resulting in clonal sectors with altered pigmentation patterns. The more common non-clonal patterns cannot be simply understood in terms of cell lineages and mitosis and include cases where an apparently uniform cell population or plant organ may contain a number of different pigmented regions. The flower species in which the determinants of the spatial patterns of pigment biosynthesis are most clearly understood is *Antirrhinum* (Coen et al., 1988). The transposable element Tam3 enabled Martin et al. (1985) to elucidate the mechanisms determining the *pallida*^{recurrens}-2 (*pal*^{rec}-2) clonal phenotype of red spots and sectors on an ivory background. The *pal*^{rec}-2 allele, encoding the DFR gene, contains within that gene the Tam3 transposon sequence, which renders the DFR gene non-functional. During flower development the Tam3 sequences can excise from the *pal*^{rec}-2 allele, restore the gene to a functional state, and result in biosynthesis of the normal red pigment. Once the Tam3 sequences are excised the restored Pal^{+} gene is clonally inherited through subsequent cell lineages resulting in red pigmented spots or sectors in the flower.

Non-clonal pigment patterns have also been studied at the molecular level in *Antirrhinum*, again utilizing mutations in the *pallida* and *nivea* loci (Spiribille and Forkman, 1982), as well as the regulatory locus *delila* (Coen et al., 1988). The *delila* mutation, when homozygous, results in a complete loss of anthocyanin pigment from the lobe of the flower. Analysis of the *pallida* and *nivea* structural gene mRNA levels in the wild type and *delila* mutant flowers indicated a 5–10 fold reduction in mRNA for the *nivea* gene and almost complete absence of *pallida* gene mRNA in the non-pigmented lobe region of the flower. Feeding experiments and assays of other enzymes in the pigment pathway indicated that genes other than *pallida* and *nivea* may also be inactive in the *delila* mutants (Coen et al., 1988), suggesting that the *delila* mutation is somehow acting in *trans* to effect the expression of a number of the genes in the pigment biosynthesis pathway. The *delila* mutation also effects anthocyanin pigment biosynthesis in other normally pigmented structures such as the stem and leaves, making this mutation analogous to the regulatory mutations which occur in maize (Dooner, 1983), and the *an*4 mutations which alter spatial patterns of pigment biosynthesis in the corolla of petunia (Mol et al., 1983). All of these regulatory mutations appear to act in *trans*, and may alter the expression of a number of structural genes resulting in different patterns of gene expression within a single region or organ of the plant.

Novel pigmentation patterns have been observed among antisense CHS transgenic petunia plants (van der Krol et al., 1988). These patterns are thought to be due to the relative antisense to sense CHS gene expression patterns; which in turn could be due to slight variations in internal physiological conditions or growth conditions during flower development (van der Krol et al., 1990b). Little is known about the genetic stability of the new flower pigmentation patterns resulting from antisense CHS gene transformations, and the alteration of pigment patterns within individual plants may present problems for commercial extension of the use of antisense gene inhibition to produce new and novel pigment patterns. Careful selection of regulatory sequences for antisense gene constructs and extensive analysis of transgenic plants may overcome the variability and stability problems presently observed.

Quite spectacular and unexpected alterations in pigment biosynthesis patterns have recently been observed in transgenic petunia over-expressing gene constructs containing the sense CHS gene (Napoli et al., 1990; van der Krol et al., 1990a) and the sense DFR gene (van der Krol et al., 1990a). In experiments investigating the effects of over-expressing pigment biosynthesis genes, these two groups found the presence of introduced sense cDNA sequences or genomic fragments containing CHS often resulted in inhibition of pigment biosynthesis. In areas of the flower where pigment production was blocked, the level of expression of both the endogenous genes (either CHS or DFR) and the introduced sense gene were dramatically suppressed. The exact mechanism of this suppression is not known, but it is suspected that the initial expression of the transgene may be required to suppress both the endogenous and transgene (van der Krol et al., 1990b).

Consideral variability in the patterns of pigments were observed, with two main patterns, a "wedge" shape and a "radial" pattern. In both cases the white sectors appeared non-clonal in origin (Napoli et al., 1990). In some cases stable ectopic expression patterns were observed, however, considerable variability of pigment patterns was observed among the progeny of the primary transgenic plants. Amongst the progeny plants containing the sense transgene there appeared to be variability in the penetrance of the phenotype, again this may be due to the region in the genome where the transgene has integrated. Whilst the production of novel patterns of pigment biosynthesis have been created by the over-expression of flavonoid pigment biosynthesis genes, the genetic stability of these patterns remains to be evaluated. It is interesting to note that the pigment patterns produced by sense expression of pigment biosynthesis genes are similar to naturally occurring petunia mutants such as Red Star (Mol et al., 1983) which also show non-clonal pigmented and non-pigmented sectors. These naturally occurring varieties also show variability in pigment patterns, being influenced by environmental conditions such as temperature and light intensities (Mol et al., 1983).

IV. Conclusion

One of the great potentials of molecular breeding lies in the alteration of not only flower colour, but also to the manipulation of other traits in species which are commercially important to the cut flower industry.

Whilst the ability to modify pigment biosynthesis and create novel pigmentation patterns by genetic engineering has clearly been demonstrated in petunia and tobacco, neither of these species is cut a flower species of great commercial value.

Of the seven most important commercial cut flower species, tulip, rose, carnation, chrysanthemum, freesia, gerbera, and daffodil, the only species for which foreign gene expression has been reported is chrysanthemum (Lemieux, 1989). The application of genetic engineering is therefore limited at present by the need to develop gene transfer systems for these commercial crops. Tulip, freesia, iris, lily, and daffodil are all monocots and by analogy with other monocots are therefore likely to be difficult to transform using *Agrobacterium tumefaciens*. Several of these species can be readily regenerated and as such may be more amenable to other transformation systems, for example microprojectile bombardment. It is in the area of transformation where some of the major hurdles to the commercial application of genetic engineering or "molecular breeding" to the cut flower industry lie. Given the rate at which pigment biosynthesis genes are being cloned and the growing commercial interest in the development of gene transfer systems for cut flower crops, the future of molecular flower breeding appears bright.

Acknowledgement

The author would like to thank Filippa Kovacic, Timothy Holton, Terese Wardley, Kim Stevenson, John Menting, and Craig Hyland for their contributions to work described in this chapter. I also wish to thank Vicki Georgiou for typing the manuscript.

Support for this work was provided under the Generic Technology component of the Industry Research and Development Act, 1986.

V. References

Asen S, Stewart RN, Norris KH (1972) Copigmentation of anthocyanins in plant tissues and its effect on colour. Phytochemistry 11: 1139–1144

Beld M, Martin C, Huits H, Stuitje AR, Gerats AGM (1989) Flavonoid synthesis in *Petunia hybrida*: partial characterization of dihydroflavonol-4-reductase genes. Plant Mol Biol 13: 491–502

Bonas U, Sommer H, Saedler H (1984) The 17 kb Tam1 element of *Antirrhinum majus* induces a 3 bp duplication upon integration into the chalcone synthase gene. EMBO J 3: 1015–1019

Chandler VL, Radicella JP, Robbins TP, Chen J, Turks D (1989) Two regulatory genes of the maize anthocyanin pathway are homologues: isolation of B utilizing R genomic sequences. Plant Cell 1: 1175–1183

Coen ES, Almeida J, Robbins TP, Hudson A, Carpenter R (1988) Molecular analysis of genes determining spatial patterns in *Antirrhinum majus*. In: Verma DPS, Goldberg RB (eds) Temporal and spatial regulation of plant genes. Springer, Wien New York, pp 63–82 [Dennis ES et al (eds) Plant gene research. Basic knowledge and application]

Coen ES, Carpenter R, Martin C (1986) Transposable elements generate novel spatial patterns of gene expression in *Antirrhinum majus*. Cell 47: 285–296

Dooner HK (1983) Co-ordinate genetic regulation of flavonoid biosynthetic enzymes in maize. Mol Gen Genet 189: 136–141

Dooner HK, Nelson OE Jr (1977) Genetic control of UDP glucose: flavonol 3-O-glucosyltransferase in the endosperm of maize. Biochem Genet 15: 509–519

Douglas C, Hoffman H, Schulz W, Hahlbrock K (1987) Structure and elicitor or UV light stimulated expression of two 4-coumarate: CoA ligase genes in parsley. EMBO J 6: 1189–1195

Edwards K, Cramer CL, Bolwell GP, Dixon RA, Schuch W, Lamb CJ (1985) Rapid transient induction of phenylalanine ammonia lyase mRNA in elicitor-treated bean cells. Proc Natl Acad Sci USA 82: 6731–6735

Fedoroff N (1983) Controlling elements in maize. In: Shapiro JA (ed) Mobile genetic elements. Academic Press, New York, pp 1–63

Fedoroff N, Furtek DB, Nelson OE Jr (1984) Cloning of the *bronze* locus in maize by a simple and generalized procedure using the transposable controlling element *Activator* (*Ac*). Proc Natl Acad Sci USA 81: 3825–3829

Feinbaum RL, Ausubel FM (1988) Transcriptional regulation of the *Arabidopsis thaliana* chalcone synthase gene. Mol Cell Biol 8: 1985–1992

Forkmann G, Ruhnau B (1987) Distinct substrate specificity of dihydroflavonol-4-reductase from flowers in *Petunia hybrida*. Z Naturforsch 42c: 1146–1148

Gabriac B, Benveniste I, Durst F (1985) Isolation and characterization of cytochrome P-450 from higher plants (*Helianthis tuberosis*). CR Acad Sci Ser III 301: 753–758

Gerats AGM, de Vlaming P, Doodeman M, Al B, Schram AW (1982) Genetic control of conversion of dihydroflavonols into flavonols and anthocyanins in flowers of *Petunia hybrida*. Planta 155: 364–368

Goto T (1987) Structure, stability and color variation of natural anthocyanins. In: Herz W et al (eds) Progress in the chemistry of organic natural products, vol 52. Springer, Wien New York, pp 113–158

Hagmann M, Heller W, Grisebach H (1983) Induction and characterization of a microsomal flavonoid 3′-hydroxylase from parsley cell cultures. Eur J Biochem 134: 547–554

Harbourne JB (1967) Comparative biochemistry of the flavonoids. Academic Press, London, pp 1–99

Heller E, Hahlbrock K (1980) Highly purified "flavonone synthase" from parsley catalyzes the formation of naringenin chalcone. Arch Biochem Biophys 200: 617–619

Herrmann A, Schulz W, Hahlbrock K (1988) Two alleles of the single-copy chalcone synthase gene in parsley differ by a transposon-like element. Mol Gen Genet 212: 93–98

Higashi K, Ikeuchi K, Obara M, Karasaki Y, Hirano H, Gotoh S, Koga Y (1985) Purification of a single form of microsomal cytochrome P-450 from tulip bulbs (*Tulip genseriana* L.). Agricult Biol Chem 49: 2399–2405

Horsch RB, Fry JE, Hoffman NL, Wallroth M, Eichholtz D, Rogers SG, Fraley RT (1985) A simple and general method for transferring genes into plants. Science 227: 1229–1231

Koes RE, Spelt CE, Reif HJ, van den Elzen PJM, Veltkamp E, Mol JNM (1986) Floral tissue of *Petunia hybrida* (V30) expresses only one member of the chalcone synthase multigene family. Nucleic Acids Res 14: 5229–5239

Kreuzaler F, Ragg H, Fautz E, Kuhn DN, Hahlbrock K (1983) UV-induction of chalcone synthase mRNA in cell suspension cultures of *Petroselinum hortense*. Proc Natl Acad Sci USA 80: 2591–2593

Kuhn D, Chappell J, Boudet A, Hahlbrock K (1984) Induction of phenylalanine ammonia lyase and 4-coumarate: CoA ligase mRNAs in cultured plant cells by UV light or fungal elicitor. Proc Natl Acad Sci USA 81: 1102–1106

Larson RL, Bussard JB (1986) Microsomal flavonoid 3′ mono oxygenase from maize seedlings. Plant Physiol 80: 483–486

Lemieux C (1989) Transformation of chrysanthemum cultivars with *Agrobacterium tumefaciens*. In: Proceedings Horticultural Biotechnology Symposium, University of California, Davis, August, 1989

Ludwig SR, Ledare HF, Dellaporta SL, Wessler SR (1989) *Lc*, a member of the maize *R* gene family responsible for tissue specific anthocyanin production encodes a protein similar to transcriptional activators and contains the *myc* homology region. Proc Natl Acad Sci USA 86: 7092–7096

Martin C, Carpenter R, Sommer H, Saedler H, Coen E (1985) Molecular analysis of instability in flower pigmentation of *Antirrhinum majus*, following isolation of the *pallida* locus by transposon tagging. EMBO J 4: 1625–1630

Matile Ph (1975) The lytic compartment of plant cells. Springer, Wien New York [Alfert M et al (eds) Cell biology monographs, vol 1]

McLaughlin M, Walbot V (1987) Cloning of mutable bz 2 allele of maize by transposon tagging and differential hybridization. Genetics 117: 771–784

Mehdy MC, Lamb CJ (1987) Chalcone isomerase cDNA cloning and mRNA induction by fungal elicitor, wounding and infection. EMBO J 6: 1527–1533

Meyer P, Heidmann I, Forkmann G, Saedler H (1987) A new petunia flower colour generated by transformation of a mutant with a maize gene. Nature 330: 677–678

Mol JNM, Schram AW, de Vlaming P, Gerats AGM, Kreuzaler F, Hahlbrock K, Reif HJ, Veltkamp E (1983) Regulation of flavonoid gene expression in *Petunia hybrida*: description and partial characterization of a conditional mutant in chalcone synthase gene expression. Mol Gen Genet 192: 424–429

Napoli C, Lemieux C, Jorgensen R (1990) Introduction of chimeric chalcone synthase gene into petunia results in reversible co-suppression of homologous genes in *trans*. Plant Cell 2: 279–289

Nevers P, Shephard NS, Saedler H (1986) Plant transposable elements. Adv Bot Res 12: 103–203

Niesbach-Klosgen K, Barzen E, Bernhardt J, Rohde W, Schwarz-Sommer Z, Reif HJ, Wienand U, Saedler H (1987) Chalcone synthase genes in plants: a tool to study evolutionary relationships. J Mol Evol 26: 213–225

O'Keefe DP, Leto KJ (1989) Cytochrome P-450 from the mesocarp of avocado (*Persea americana*). Plant Physiol 89: 1141–1149

O'Reilly C, Shephard N, Pereira A, Schwarz-Sommer Z, Bertram I, Robertson DS, Peterson PA, Saedler H (1985) Molecular cloning of the A1 locus of *Zea mays* using the transposable elements En and Mu 1. EMBO J 4: 877–882

Paz-Ares J, Weinand U, Peterson PA, Saedler H (1986) Molecular cloning of the c locus of *Zea mays*: a locus regulating the anthocyanin pathway. EMBO J 5: 829–833

Paz-Ares J, Ghosal D, Wienand U, Peterson PA, Saedler H (1987) The regulatory C1 locus of *Zea mays* encodes a protein with homology to *myb* proto-oncogene products with structural similarities to transcription activators. EMBO J 6: 3553–3558

Pereira A, Cuypers H, Gierl A, Schwarz-Sommer Z, Saedler H (1986) Molecular analysis of the En/Spm transposable element system of *Zea mays*. EMBO J 5: 835–841

Pereira A, Schwarz-Sommer Z, Gierl A, Bertram I, Peterson PA Saedler H (1985) Genetic and molecular analysis of Enhancer (En) transposable element system of *Zea mays*. EMBO J 4: 17–23

Ranjeva R, Boudet AM, Faggion R (1976) Phenolic metabolism in petunia tissues IV. Properties of p-coumarate: coenzyme A ligase isoenzymes. Biochemie 58: 1255–1262

Reddy A, Britsch L, Salamini F, Saedler H, Rohde W (1987) The A1 (anthocyanin-1) locus in *Zea mays* encodes dihydroquercetin reductase. Plant Sci 52: 7–13

Robinson GM, Robinson R (1931) A survey of anthocyanins I. Biochem J 25: 1687–1705

Russell DW (1971) The metabolism of aromatic compounds in higher plants. J Biol Chem 246: 3870–3878

Ryder TB, Hedrick SA, Bell JN, Liang X, Clouse SD, Lamb CJ (1987) Organization and differential activation of a gene family encoding the plant defense enzyme chalcone synthase in *Phaseolus vulgaris*. Mol Gen Genet 210: 219–233

Schroder J, Kruezaler F, Schafer E, Hahlbrock K (1979) Concomitant induction of phenylalanine ammonia-lyase and flavanone synthase mRNA's in irradiated plant cells. J Biol Chem 254: 57–65

Schwarz-Sommer Z, Shephard N, Tacke E, Gierl A, Rohde W, Leclercq L, Mattes M, Berndtgen R, Peterson P, Saedler H (1987) Influence of transposable elements on the structure and function of the A1 gene of *Zea mays*. EMBO J 6: 287–294

Sommer H, Saedler H (1986) Structure of the chalcone synthase gene of *Antirrhinum majus*. Mol Gen Genet 202: 429–434

Sommer H, Carpenter R, Harrison BJ, Saedler H (1985) The transposable element Tam3 of *Antirrhinum majus* generates a novel type of sequence alteration upon excision. Mol Gen Genet 199: 225–231

Spribille R, Forkmann G (1982) Chalcone synthesis and hydroxylation of flavonoids in 3′-position with enzyme preparations from flowers of *Dianthus caryophyllus* L. (carnation). Planta 155: 176–182

Stotz G, Forkmann G (1982) Hydroxylation of the B-ring of flavonoids in the 3′ and 5′ position with enzyme extracts from flowers of *Verbena hybrida*. Z Naturforsch 37c: 19–32

Stotz G, de Vlaming P, Wiering H, Schram AW, Forkmann G (1985) Genetic and biochemical studies on flavonoid 3′ hydroxylation in flowers of *Petunia hybrida*. Theor Appl Genet 70: 300–305

Strickland RG, Harrison BJ (1974) Precursors and genetic control of pigmentation. I Induced biosynthesis of pelargonidin, cyanidin and delphinidin in *Antirrhinum majus*. Heredity 33: 108–112

Upadhyaya KC, Sommer H, Krebbers E, Saedler H (1985) The paramutagenic line *niv*-44 has a 5 kb insert, Tam2, in the chalcone synthase gene of *Antirrhinum majus*. Mol Gen Genet 199: 201–207

van der Krol AR, Lenting PE, Veenstra J, van der Meer IM, Koes RE, Gerats AGM, Mol JNM, Stuitje AR (1988) An antisense chalcone synthase gene in transgenic plants inhibits flower pigmentation. Nature 333: 866–869

van der Krol AR, Mur LA, Beld M, Mol JNM, Stuitje AR (1990a) Flavonoid genes in petunia: addition of a limited number of gene copies may lead to a suppression of gene expression. Plant Cell 2: 291–299

van der Krol AR, Mur LA, de Lange P, Gerats AGM, Mol JNM, Stuitje AR (1990b) Antisense chalcone synthase genes in petunia: visualization of variable transgene expression. Mol Gen Genet 220: 204–212

van Tunen A, Koes RE, Spelt CE, van der Krol AR, Stuitje AR, Mol JNM (1988) Cloning of the two chalcone flavanone isomerase genes from *Petunia hybrida*: co-ordinate, light-regulated and differential expression of flavonoid genes. EMBO J 7: 1257–1263

Wiering H (1974) Genetics of flower colour in *Petunia hybrida*. Horticult Genen Phaenen 17: 117–134

Wiering H, deVlaming P (1984) Inheritance and biochemistry of pigments. In: Sink KC (ed) Petunia. Springer, Berlin Heidelberg New York Tokyo, pp 49–67 [Frankel R et al (eds) Monographs on theoretical and applied genetics, vol 9]

Wienand U, Weydemann U, Niesbach-Klosgen U, Peterson PA, Saedler H (1986) Molecular cloning of the C2 locus from *Zea mays*, the gene coding for chalcone synthase. Mol Gen Genet 203: 202–207

Chapter 8

Molecular Aspects of Self-incompatibility in the Solanaceae

Volker Haring, Bruce A. McClure, and Adrienne E. Clarke

Plant Cell Biology Research Centre, School of Botany, University of Melbourne, Parkville, Victoria 3052, Australia

Contents

I. Introduction

Many species of flowering plants have evolved genetically controlled mechanisms to prevent inbreeding. Fertilization in flowering plants involves a complex series of interactions between the haploid pollen which contains the male gametes, and the diploid tissues of the female pistil. Flowers are often hermaphroditic bearing the male and female organs in close proximity, so that the effectiveness of these mechanisms in preventing self-fertilization and promoting outcrossing is particularly important. Self-incompatibility (SI) is one of the most widespread mechanisms for preventing self-fertilization. It is the inability of seed plants to produce viable seeds after self-pollination. Study of the mechanism of self-incompatibility is not only of interest in relation to pollination but also as a model system for understanding cell–cell recognition in higher plants.

The process of fertilization and the phenomenon of self-incompatibility have fascinated biologists since Darwin's time. There is an extensive early literature describing self-incompatibility for a range of species in terms of classical genetics and microscopic observations. Much of this has been presented by de Nettancourt (1977) in his classical monograph "Incompatibility in Angiosperms" and recent developments have been reviewed on several occasions (Gibbs, 1986; Nasrallah and Nasrallah, 1986; Charlesworth, 1988; Cornish et al., 1988a, b; Ebert et al., 1989; Pettitt et al., 1989).

Since 1985, recombinant DNA technology has been applied to analyze the nature and control of the genes responsible for the phenomenon of SI. This approach resulted in the isolation and characterization of cDNAs encoding *S*-gene products from a number of species (Nasrallah et al., 1985, 1987; Anderson et al., 1986, 1989; Ai et al., 1990; Kheyr-Pour et al., 1990; Sims, pers. comm.; Thompson, pers. comm.). Recently, the first report on an enzymatic function for the *S*-glycoproteins of *Nicotiana alata* has been published; McClure et al. (1989) presented evidence that the isolated *S*-glycoproteins of *N. alata* are ribonucleases.

In this review, we give a brief description of the biology and molecular genetics of self-incompatibility in Solanaceae and then concentrate on the implications of the finding that *S*-gene products are ribonucleases.

II. Biology and Genetics of Self-incompatibility in the Solanaceae

The pollen grains from solanaceous plants contain two cells at anthesis, the vegetative cell which controls growth of the pollen tube and the generative cell which divides by mitosis after germination. In a compatible mating, the pollen hydrates and germinates on the stigma surface, the developing pollen tube penetrates the stigma and grows extracellularly through the transmitting tissue of the style to the ovary. The two sperm cells and the vegetative nucleus are carried in the tip of the pollen tube. At regular intervals along the length of the pollen tube, callose which stains with the aniline blue fluorochrome is deposited (Stone et al., 1984); this gives the pollen tube a ladder-like appearance. These deposits are believed to cut off the cytoplasm in the tip of the pollen tube from the spent pollen grain. In the ovary, the two sperm cells enter the embryo sac to complete the double fertilization characteristic of angiosperms.

Incompatible pollen tubes initially appear similar to compatible ones. However, at some point during their growth, the callose deposits become irregular and growth is inhibited. The tube walls stain more intensely with the aniline blue fluorochrome than those of compatible tubes, and there is often a characteristic deposit of callose immediately behind the tip. In some cases the tips of incompatible tubes may burst.

In the Solanaceae, self-incompatibility is controlled by a single gene, the *S*-gene, with multiple alleles. The incompatibility phenotype of a pollen grain is determined by its own haploid genotype; pollen tube growth is arrested if the *S*-allele carried by the pollen is identical to one of the two *S*-

alleles of the female tissue. Since expression of the *S*-gene in the haploid pollen is the primary determinant of the *S*-phenotype, this system is referred to as the gametophytic system. It is the most common type of self-incompatiblity and occurs in nearly half of the families of flowering plants including Solanaceae, Liliaceae, Poaceae, Commelinaceae, Onagraceae, Papaveraceae, Rosaceae, and Rubiaceae. However, it has been studied in most detail in members of the Solanaceae. The other major type of homomorphic self-incompatibility is the sporophytic system in which the phenotype of the pollen is determined by the phenotype of the pollen producing plant. This system is less widely distributed and has only been described for six families (Charlesworth, 1988). It has been studied in most detail for *Brassica* spp.

A theory to account for the two types of self-incompatibility was proposed by Lewis (1949, 1960), Pandey (1958, 1960), and Heslop-Harrison (1975). It suggests that in the gametophytic system, *S*-gene activation in pollen occurs after completion of meiosis. Thus Mendelian segregation results in two microspores that express one *S*-allele and two that express the other. In contrast, in the sporophytic system, the *S*-genes are thought to be activated before completion of meiosis so that *both* parental *S*-alleles are expressed in all four microspores.

Self-incompatibility in the Solanaceae is developmentally regulated. That is, it is not expressed in immature flowers, but is expressed in the mature flower. At early developmental stages, for example when the bud is green, the flower can be opened and the immature style pollinated with pollen from a mature flower. This treatment can lead to viable seeds and is used to produce plants homozygous for particular *S*-alleles. The genetics and biology of self-incompatibility have been comprehensively reviewed by Lewis (1960), Pandey (1968), Linskens (1975), de Nettancourt (1977), Cornish et al. (1988a), and Pettitt et al. (1989).

III. Molecular Biology of Self-incompatibility

A variety of techniques have been used to identify *S*-allele associated molecules with the ultimate aim of determining the molecular mechanisms of self-incompatibility. In the early 1950s, Lewis (1952) pioneered the application of serological techniques to study self-incompatibility and was the first to detect *S*-allele related substances using pollen of *Oenothera organensis* as his experimental system. This was followed by the observation that certain style components apparently correspond to particular *S*-genotypes in *Petunia hybrida* (Linskens, 1960). The first convincing evidence that a pistil protein corresponded to a particular *S*-allele was presented by Hinata and Nishio (1978). Using electrophoretic techniques, they showed that a stigma protein corresponding to a particular *S*-allele of *Brassica campestris* segregated with this allele through the F_2 generation, indicating that the *S*-related proteins are indeed products of the *S*-alleles or of a closely related gene. Subsequently, stigma or style components were correlated with individual *S*-alleles for several systems (reviewed, in Nasrallah and Nasrallah, 1986; Cornish et al., 1988b). Most of the *S*-related components analyzed so far are basic proteins which are glycosylated and are referred to as *S*-glycoproteins. Their size ranges differ significantly between sporophytic (55–65 kDa) and gametophytic systems (24–35 kDa). Recent progress towards understanding the molecular basis of self-incompatibility was stimulated by the cloning of cDNAs corresponding to *S*-locus specific glycoproteins (Nasrallah et al., 1985; Anderson et al., 1986).

cDNAs corresponding to *S*-glycoproteins have now been cloned from a number of related species, and sequence data has been reported for some of them (Nasrallah et al., 1985, 1987; Anderson et al., 1986, 1989; Ai et al., 1990; Kheyr-Pour et al., 1990). There are several lines of evidence that these cDNAs and *S*-glycoproteins are derived from the *S*-locus.

In *N. alata*, hybridization analysis using cDNAs as probes on Southern blots containing genomic DNA restriction fragments is consistent with a single locus, single copy gene system as proposed on the basis of classical genetics (Bernatzky et al., 1988). Further, restriction fragment length polymorphisms characteristic of particular *S*-alleles were also abundant. The style glycoproteins can be separated by FPLC and give elution profiles characteristic of each allele (Jahnen et al., 1989a). These glycoproteins segregate with particular *S*-alleles.

Further, the amino acid sequences of these glyproteins correspond to the deduced sequences from the cloned cDNAs. The expression of the *S*-glycoproteins is spatially and temporally regulated in a manner consistent with their role in self-incompatibility, as they are expressed only in the mature style and not in the immature style when self-incompatibility is not functional (Anderson et al., 1986). In situ hybridization of cDNA to style sections mirrors the developmental and tissue specific expression of the *S*-glycoproteins (Cornish et al., 1987). Finally, the purified *S*-glycoproteins from *N. alata* inhibit pollen tube growth in vitro, with some allelic specificity (Jahnen et al., 1989b). Comparison of the amino acid sequences derived from the cloned cDNAs corresponding to *S*-alleles from several solanaceous species shows that the *S*-glycoproteins are related (Table 1). The most homologous group are the sequences of *P. inflata* with an average homology of 75.6% (Ai et al., 1990) compared to 53.8% of the *N. alata* group (Kheyr-Pour et al., 1990). Interestingly, the *N. alata* alleles S_Z and S_{F11} seem to be more closely related to the *Petunia S*-alleles than to the other *N. alata S*-alleles. Both groups of sequences show defined regions of homology and variability (Anderson et al., 1989; Kheyr-Pour et al., 1990; Ai et al., 1990). The variable regions could harbour allelic specificity and conserved regions could define domains involved in other self-incompatibility functions and in

Table 1. Scores for alignments of *S*-glycoproteins from *N. alata*, *P. inflata*, and *S. tuberosum* as well as the two ribonucleases T_2 and Rh

	$N\text{–}S_2$	$N\text{–}S_3$	$N\text{–}S_6$	$N\text{–}S_{F11}$	$N\text{–}S_Z$	$P\text{–}S_1$	$P\text{–}S_2$	$P\text{–}S_3$	$S\text{–}S_1$	$A\text{–}T_2$	R–Rh
$N\text{–}S_1$	40	35	39	30	26	26	28	31	27	8	10
$N\text{–}S_2$		33	40	31	30	27	26	27	31	8	10
$N\text{–}S_3$			34	25	26	21	23	23	28	8	12
$N\text{–}S_6$				27	29	29	26	29	26	8	10
$N\text{–}S_{F11}$					34	30	33	37	33	8	8
$N\text{–}S_Z$						41	43	41	30	8	7
$P\text{–}S_1$							43	44	29	5	6
$P\text{–}S_2$								44	30	6	7
$P\text{–}S_3$									28	6	7
$S\text{–}S_1$										6	7
$A\text{–}T_2$											23

Alignment scores for comparison of ten *S*-glycoproteins and two fungal ribonucleases generated by the program ALIGN (Dayhoff et al., 1983). The scores are units of standard deviations from the mean score of 100 alignments using randomly permutated sequences. They were determined with the mutation data matrix (250 PAMs) with a bias of +6 and a gap penalty of 9. Scores greater than 10 are assumed to be highly significant. The sequences are from N, *Nicotiana alata* (Anderson et al., 1989; Kheyr-Pour et al., 1990); P, *Petunia inflata* (Ai et al., 1990); S, *Solanum tuberosum* (Thompson, pers. comm.), A, *Aspergillus oryzae* (Kawata et al., 1988), and R, *Rhizopus niveus* (Horiuchi et al., 1988)

maintaining the overall structure (as for example the highly conserved cysteine residues). However, a dissection of the *S*-sequences into functional domains requires further data.

IV. The Style *S*-glycoproteins of *Nicotiana alata* Are Ribonucleases

Two of the highly conserved regions in the sequences of *N. alata* S_2-, S_3-, and S_6-glycoproteins show extended similarities with sequences of two fungal ribonucleases, RNase T_2 (Kawata et al., 1988) and RNase Rh (Horiuchi et al., 1988). These regions of homology include two histidine residues which have been implicated in RNase T_2 catalysis (Kawata et al., 1988). Moreover, four cysteine residues around these putative active site histidine residues are also conserved. These findings suggest that the *S*-glycoproteins share some structural similarities with the fungal ribonucleases and could themselves be ribonucleases. Indeed, McClure et al. (1989) found high levels of RNase activity in extracts of *N. alata* styles but not in extracts of styles of *N. tabacum* which is self-compatible and does not produce an *S*-like style glycoprotein (Jahnen et al., 1989a). Furthermore, cation exchange chromatography of style extracts from four alleles (S_1, S_3, S_6, and S_7) showed that the *S*-glycoproteins coeluted with the major part of ribonuclease activity of style extracts. During this chromatographic procedure, the *S*-glycoproteins separate as the major peaks in the elution profile, and the position in the profile is characteristic for a particular *S*-glycoprotein (Jahnen et al., 1989a). The coincidence of RNase activity and *S*-glycoprotein peaks in the profiles of different *S*-alleles demonstrates that the RNase activity is an inherent component of the *S*-glycoproteins. Thus the *S*-glycoproteins are also be referred to as *S*-RNases.

The specific activity of the *S*-RNases in style extracts of homozygous plants is comparable to that of RNase T_2 but varies about tenfold between different *S*-alleles. This variation does not correlate with the relative amount of the particular *S*-glycoproteins in the style and therefore may reflect a variation in the specific activity of the individual *S*-glycoproteins.

The nature of gametophytic self-incompatibility suggests that the *S*-gene product must be expressed in the pollen as well as in the style. However, McClure et al. (1989) found only trace amounts of RNase activity in extracts of in vitro germinated pollen indicating that the *S*-RNase is not highly expressed in pollen.

This finding raises the question of whether the ribonuclease activity is directly involved in the arrest of pollen tube growth. The observation that style extracts from the self-compatible species *N. tabacum* (McClure et al., 1989) and a self-compatible accession of *Lycopersicum peruvianum* (Lewis and McClure, unpubl. results) neither contain an *S*-glycoprotein-like component nor an *S*-RNase-like activity suggests that it may well be. It is, however, possible that during evolution of self-incompatibility, plants have modified an already existing protein (i.e., RNase) to fulfil a signalling function in self-incompatibility without loss of its enzymatic function. A classic example of this kind of recruitment of enzymes for a non-enzymatic function is the evolution of crystallins, the structural proteins of the eye lens (reviewed, in deJong et al., 1989). Several crystallins are similar, or in some cases identical, to common enzymes, but some have lost most or all of their former activity. At least one, δ-crystallin in chicken, seems to have arisen from gene duplication, leading to loss of enzymatic activity in the lens protein. For others, e.g., τ-crystallin of ducks, only a single gene is used to express the crystallin in the lens and the enzyme in other tissues; this phenomenon is described as gene sharing.

Following this analogy, an ancestral RNase might have been recruited for a role in self-incompatibility; this role may not have required conservation of the enzymatic activity per se, but rather required some other feature of the ancestral protein such as its structure, pattern of expression or stability to the extracellular environment. However, as the *S*-glycoproteins and the RNase activity are lacking in the styles of self-compatible species, the elevated style RNase-activity, characteristic of self-incompatible *N.*

alata and *L. peruvianum* accessions, is not vital for the survival of the plant. The conservation of the active site histidines in the *S*-glycoproteins of *Petunia inflata* and *Solanum tuberosum* suggests that these proteins probably also have RNase-activity. A chance conservation of this activity during "recruitment" of the RNase for a self-incompatibility function seems unlikely, since only some amino acids near the active site histidines are highly conserved in an otherwise hypervariable region (Anderson et al., 1989; Kheyr-Pour et al., 1990). This suggests that these residues have been conserved for the maintenance of an RNase function. The observation that a non-homologous RNase can inhibit pollen tube growth in vitro (McClure et al., 1989) lends further weight to the hypothesis that the RNase activity is involved in self-incompatibility function.

V. Function of RNA Degrading Enzymes in Cell–Cell Interactions in Other Systems

There are several systems in which RNases are implicated in cell–cell communication. These are now examined as a basis for developing ideas as to how the style *S*-RNases may function in self-incompatibility.

A. Plant Cytotoxins Degrading RNA

Many plants encode highly cytotoxic proteins (Stirpe and Barbieri, 1986) that hydrolyse a specific N-glycosidic bond in 28S RNA of eukaryotic ribosomes (Endo et al., 1988). This modification renders the ribosome inactive. These plant cytotoxins, known as ribosome inactivating proteins (RIPs), can be divided into two groups. One group, exemplified by pokeweed antiviral protein, contains those RIPs that consist of a single polypeptide chain. These proteins are potent inhibitors of protein synthesis in cell-free systems but are relatively non-toxic to intact cells. The RIPs of the other group, which includes ricin and abrin, consist of two non-identical subunits, the A- and B-chains. The A-chains are similar to the polypeptides of the former group and act by the same mechanism. The B-chain is a lectin that binds to cell surface glycoproteins and facilitates the uptake of the A-chain by the cell, making this group highly toxic to intact cells.

This system of uptake is relevant to thinking about how the style RNases might be taken up by the pollen tube. It may be that allelic specificity is exerted during uptake of *S*-RNase. This would require that a certain *S*-glycoprotein is only taken up by pollen tubes of the same *S*-genotype. In this case, the proposed pollen *S*-product could exert a specific uptake function analogous to the B-chains of the RIPs. This mechanism requires that the pollen-encoded function specifically interacts with the style *S*-RNase of its own genotype. The observation that in vitro growth of *N. alata* pollen is inhibited by exogenously supplied bovine pancreatic RNase A (Lush, unpubl. observation) suggests that enzymes can penetrate the pollen tube in an active form independent of a specific uptake pathway. Such an unbiased susceptibility of the pollen tubes for the *S*-RNases might also account for the in vivo inhibition of growth of *N. tabacum* pollen tubes by all *S*-genotypes of *N. alata*. Therefore, it seems unlikely that the uptake mechanism is the major site of specific interactions between pollen and style *S*-allele products.

B. The Bacterial Colicins

This is an example of a mechanism for specific inhibition of nucleases. Some bacterial strains produce plasmid-encoded proteins which are toxic to strains other than the producer (reviewed, in Koniski, 1982). The immunity of the producer strain is due to another plasmid-encoded peptide which interacts specifically with one type of bacteriocin but not with other closely related types. Bacteriocins with nucleolytic activity are the colicins E2, E3, E8, and cloacin DF13. Colicins E2 and E8 are DNA endonucleases whereas colicin E3 and cloacin DF13 are highly specific RNases. The latter are functionally similar to the plant RIPs, in that they inactivate bacterial ribosomes by a specific cleavage of the 16S rRNA. The structure of these colicins can be delineated into three functional domains. The N-terminal part, comprising about 25% of the 50–60 kDa polypeptides has a rather hydrophobic character and is thought to be involved in translocation of the toxins across the cell membrane. The central part interacts with specific outer membrane proteins of the target cell. As with the plant RIPs, the colicins require interaction with a receptor for transport into the target cell. The C-terminal portion of the colicins contains the catalytic domain. The corresponding immunity protein binds tightly to this region and masks the nucleolytic activity inside the producer cell. Furthermore, this binding also protects some sites in the colicins from proteolytic attack. The interaction of immunity protein and toxin is absolutely specific. For example, colicin E3 activity is only neutralized by its own immunity protein but not by that of cloacin DF13 despite the fact that the catalytic domains have 84% homology (Masaki and Ohta, 1985).

If an analogous system operates in self-incompatibility, a mechanism opposite to the isotype inhibition of the colicins would be required. This could be accomplished by pollen proteins inactivating all style *S*-RNases except that of the same *S*-allele. The pollen *S*-products would be expected to bind to the style RNases. Pollen *S*-products of all *S*-alleles, except the self-product, would need the capacity to bind. This is an example of the "oppositional model" (de Nettancourt, 1977). Alternatively, a model in which a pollen *S*-product binds the style *S*-RNase so that the enzymatic function is activated, could be envisaged. This would require binding of the pollen product of the same allele, as that encoding the style *S*-RNase. Such a mechanism could be important, if entry of style *S*-RNase into the pollen tube is limited; that is, if insufficient RNase activity enters the pollen tube to cause arrest of growth without activation. Another possibility would be that binding of a pollen *S*-gene product to the style *S*-RNase of the same allele, effectively protects this *S*-RNase from proteolytic attack. These are examples of the "complementary model".

C. RNase Activity of Animal Hormones

Recently extracellular RNases have been suggested as controlling factors in several aspects of growth and development of vertebrates (Benner and Allemann, 1989). Certain molecular factors implicated in blood vessel formation (angiogenesis) are particularly intriguing. For example, among the factors known to induce angiogenesis in vitro is a secreted polypeptide known as angiogenin. Angiogenin has been cloned and sequence analysis shows it to be a member of the pancreatic ribonuclease superfamily (Strydom et al., 1985). It has intrinsic RNase activity and it has been suggested that its enzymatic function is related to its angiogenic activity. Interestingly, human placental ribonuclease inhibitor (HPRI), an intracellular protein, has an even higher affinity for angiogenin than for pancreatic RNases (Shapiro and Vallee, 1987). Thus HPRI inhibits both the RNase activity and the in vitro angiogenic activity of extracellular angiogenin. The biological significance of this observation is not yet clear. It has been suggested that the widespread occurrence of extracellular RNases implies the existence of extracellular RNA substrates (Benner, 1988). Indeed there does appear to be precedent for the existence of extracellular RNAs associated with morphogenetic activity. Angiotropin, a factor secreted by certain tumor lines, contains both RNA and peptide components as well as a single bound copper ion (Wissler et al., 1986). It is not known what, if any, relationship this RNA containing factor has to the activity of angiogenin. Nevertheless, it has been suggested that the existence of mutually regulated RNA/RNase/inhibitor factors provides the potential for an effective regulatory triad (Benner, 1988). There is no obvious relationship of this system to the growth regulating activity of the *S*-glycoproteins but it serves to emphasize that we must not limit our search for the in vivo RNA substrate to intracellular molecules. Perhaps pollen tubes secrete such an extracellular communicator-particle composed of an RNA- and a protein-moiety. This hypothetical particle could bear an allelic specificity determinant. In this case, we envisage that some interaction between *S*-RNase and such a particle (perhaps hydrolysis of the RNA-moiety) might lead to pollen tube growth inhibition in an incompatible cross but allow continued growth of compatible pollen tubes.

VI. Conclusion

The finding that the *S*-glycoproteins in Solanaceae are RNases leads to a series of new questions and possibilities. The RNase activity has only been demonstrated for *N. alata* glycoproteins. Searches of the published sequence data from *Brassica* spp. *S*-glycoproteins do not show any obvious homology with different types of RNases (Haring, unpubl. observations). It is, at this stage, not clear whether *S*-glycoproteins from other plant families have RNase activity or activities corresponding to other cytotoxins.

In this review, we have enumerated systems in which controlled RNA degradation contributes to biological activity. We have then developed models of how analogous mechanisms could be involved in self-incompatibility. At this stage the models are hypothetical only, but are a useful framework for future thinking and experimentation.

Acknowledgements

We gratefully acknowledge the input of all our colleagues, but particularly wish to thank M. Anderson, A. Bacic, J. Gray, S.-L. Mau, G. McFadden, S. Read, and J. Woodward for their contributions and stimulating discussions. The first author is supported by a fellowship from Deutsche Forschungsgemeinschaft.

VII. References

Ai Y, Singh A, Coleman CE, Ioerger TR, Kheyr-Pour A, Kao T-H (1990) Self-incompatibility in *Petunia inflata*: isolation and characterization of cDNAs encoding three *S*-allele-associated proteins. Plant Cell 3: 130–138

Anderson MA, Cornish EC, Mau S-L, Williams EG, Hoggart R, Atkinson A, Bönig I, Grego B, Simpson R, Roche PJ, Haley JD, Niall HD, Tregear GW, Coghlan JP, Crawford RJ, Clarke AE (1986) Cloning of cDNA for a stylar glycoprotein associated with expression of self-incompatibility in *Nicotiana alata*. Nature 321: 38–44

Anderson MA, McFadden GI, Bernatzky R, Atkinson A, Orpin T, Dedman H, Tregear GW, Fernley R, Clarke AE (1989) Sequence variability of three alleles of the self-incompatibility gene of *Nicotiana alata*. Plant Cell 1: 483–491

Benner SA (1988) Extracellular 'communicator RNA'. FEBS Lett 233: 225–228

Benner SA, Allemann RK (1989) The return of pancreatic ribonucleases. TIBS 14: 396–397

Bernatzky R, Anderson MA, Clarke AE (1988) Molecular genetics of self-incompatibility in flowering plants. Dev Genet 9: 1–12

Charlesworth D (1988) Evolution of homomorphic sporophytic self-incompatibility. Heredity 60: 445–453

Cornish EC, Pettitt JM, Bönig I, Clarke AE (1987) Developmentally controlled expression of a gene associated with self-incompatibility in *Nicotiana alata*. Nature 326: 99–102

Cornish EC, Anderson MA, Clarke AE (1988a) Molecular aspects of fertilization in flowering plants. Annu Rev Cell Biol 4: 209–228

Cornish EC, Pettitt JM, Clarke AE (1988b) Self-incompatibility genes in flowering plants. In: Verma DPS, Goldberg RB (eds) Temporal and spatial regulation of plant genes. Springer, Wien New York, pp 117–130 [Dennis ES et al (eds) Plant gene research. Basic knowledge and application]

Dayhoff MO, Barker WC, Hunt LT (1983) Establishing homologies in protein sequences. Methods Enzymol 91: 524–545

de Nettancourt D (1977) Incompatibility in angiosperms. Springer, Berlin Heidelberg New York [Frankel R, Gall GAE, Linskens HF (eds) Monographs on theoretical and applied genetics, vol 3]

deJong WW, Hendriks W, Mulders JWM, Bloemendal H (1989) Evolution of eye lens crystallins: the stress connection. Trends Biol Sci 14: 365–368

Ebert PR, Anderson MA, Bernatzky R, Altschuler M, Clarke AE (1989) Genetic polymorphism of self-incompatibility in flowering plants. Cell 56: 255–262

Endo Y, Tsurugi K, Lambert JM (1988) The site of action of six different ribosome-inactivating proteins from plants on eukaryotic ribosomes: the RNA N-glycosidase activity of the proteins. Biochem Biophys Res Comm 150: 1032–1036

Gibbs P (1986) Do homomorphic and heteromorphic self-incompatibility systems have the same sporophytic mechanisms? Plant Syst Evol 154: 285–323

Heslop-Harrison J (1975) Incompatibility and the pollen–stigma interaction. Annu Rev Plant Physiol 26: 403–425

Hinata K, Nishio T (1978) *S*-allele specificity of stigma proteins in *Brassica oleraceae* and *B. campestris*. Heredity 41: 93–100

Horiuchi H, Yanai K, Takagi M, Yano K, Wakabayashi E, Sanda A, Mine S, Ohgi K, Irie M (1988) Primary structure of a base non-specific ribonuclease from *Rhizopus niveus*. J Biochem 103: 408–418

Jahnen W, Batterham MP, Clarke AE, Moritz RL, Simpson RJ (1989a) Identification, isolation, and N-terminal sequencing of style glycoproteins associated with self-incompatibility in *Nicotiana alata*. Plant Cell 1: 493–499

Jahnen W, Lush WM, Clarke AE (1989b) Inhibition of in vitro pollen tube growth by isolated *S*-glycoproteins of *Nicotiana alata*. Plant Cell 1: 501–510

Kawata Y, Sakiyama F, Tamaoki H (1988) Amino-acid sequence of ribonuclease T_2 from *Aspergillus oryzae*. Eur J Biochem 176: 683–697

Kheyr-Pour A, Bintrim SB, Ioerger TR, Remy R, Hammond SA, Kao T-H (1990) Sequence diversity of pistil *S*-proteins associated with gametophytic self-incompatibility in *Nicotiana alata*. Plant Cell 3: 88–97

Koniski J (1982) Colicins and other bacteriocins with established modes of actions. Annu Rev Microbiol 36: 125–144

Lewis D (1949) Incompatibility in flowering plants. Biol Rev 24: 472–496

Lewis D (1952) Serological reactions of pollen incompatibility substances. Proc R Soc Lond [Biol] 140: 127–135

Lewis D (1960) Genetic control of specificity and activity of the *S* antigen in plants. Proc R Soc Lond [Biol] 151: 468–477

Linskens HF (1960) Zur Frage der Entstehung der Abwehrkörper bei der Inkompabilitätsreaktion von Petunia. III. Mitteilung: Serologische Teste mit Leitsgewebs- und Pollenextrakten. Z Bot 48: 126–135

Linskens HF (1975) Incompatibility in petunia. Proc R Soc Lond [Biol] 188: 299–311

Masaki H, Ohta T (1985) Colicin E3 and its immunity genes. J Mol Biol 182: 217–227

McClure BA, Haring V, Ebert PR, Anderson MA, Simpson RJ, Sakiyama F, Clarke AE (1989) Style self-incompatibility gene products of *Nicotiana alata* are ribonucleases. Nature 342: 955–957

Nasrallah JB, Kao T-H, Goldberg ML, Nasrallah ME (1985) A cDNA clone encoding an *S* locus-specific glycoprotein from *Brassica oleraceae*. Nature 318: 263–267

Nasrallah ME, Nasrallah JB (1986) Molecular biology of self-incompatibility in plants. Trends Genet 2: 239–244

Nasrallah JB, Kao T-H, Chen C-H, Goldberg ML, Nasrallah ME (1987) Amino-acid sequence of glycoproteins encoded by three alleles of the S locus of *Brassica oleraceae*. Nature 326: 617–619

Pandey KK (1958) Time of *S*-allele action. Nature 181: 1220–1221

Pandey KK (1960) Time and site of the *S*-gene action, breeding systems and relationships in incompatibility. Euphytica 19: 364–372

Pandey KK (1968) Compatibility relationships in flowering plants: role of the *S*-gene complex. Am Naturalist 102: 475–489

Pettitt JM, Cornish EC, Clarke AE (1989) Structure and regulation of organ- and tissue-specific genes: structural and cytological features of incompatibility gene expression in flowering plants. In: Vasil IK, Schell J (eds) Cell culture and somatic cell genetics of plants. Molecular biology of plant nuclear genes, vol 6. Academic Press, New York, pp 229–261

Shapiro R, Vallee BL (1987) Human placental ribonuclease inhibitor abolishes both angiogenic and ribonucleolytic activities of angiogenin. Proc Natl Acad Sci USA 84: 2238–2241

Stirpe F, Barbieri L (1986) Ribosome-inactivating proteins up to date. FEBS Lett 195: 1–8

Stone BA, Evans NA, Bönig I, Clarke AE (1984) The application of Sirofluor, a chemically defined fluorochrome from aniline blue, for the histochemical detection of callose. Protoplasma 122: 191–195

Strydom DJ, Fett JW, Lobb RR, Alderman EM, Bethune JL, Riordan JF, Vallee BL (1985) Amino acid sequence of human tumor derived angiogenin. Biochemistry 24: 5486–5494

Wissler JH, Logemann E, Meyer HE, Krützfeld B, Höckel M, Heilmeyer Jr LMG (1986) Structure and function of a monocytic blood vessel morphogen (angiotropin) for angiogenesis in vivo: a copper-containing metallo-polyribonucleo-polypeptide as a novel and unique type of monokine. Prot Biol Fluids 34: 525–536

Subject Index

The third volume

A. D. Blonstein, P. J. King (eds.)

A Genetic Approach to Plant Biochemistry

1986. 30 figs. XI, 291 pages.
Cloth DM 128,—, öS 896,—. ISBN 3-211-81912-6

This volume brings together for the first time some interesting examples of the contributions being made by genetics to the study of plant biochemistry, including some biochemical aspects of plant development. A wide range of topics is reviewed including plant hormones, photosynthesis, nitrogen metabolism, protein synthesis, and resistance to pathogens. Two chapters deal with new methods for isolating mutants at the plant level and in protoplast culture.

The fourth volume

Th. Hohn, J. Schell (eds.)

Plant DNA Infectious Agents

1987. 76 figs. XIV, 348 pages.
Cloth DM 198,—, öS 1380,—. ISBN 3-211-81995-9

In the past few years rapid progress has been made transforming plant tissue by introducing foreign DNA. Methods make use either of viruses and soilbacteria, or involve technical manipulations of single cells followed by plant regeneration. It is now possible to spread certain genes systemically in a plant by rubbing hybrid virus nucleic acid onto a leaf and to transform its germline by infecting it with manipulated agrobacteria, thus bringing us closer to the prospect of developing new seed stocks with favourable properties such as past resistance and high nutritional value. This volume gives an account of these technologies, in addition providing basic knowledge on the strategies of natural cell invadors

The fifth volume

D. P. S. Verma, R. B. Goldberg (eds.)

Temporal and Spatial Regulation of Plant Genes

1988. 55 figs. XIII, 344 pages.
Cloth DM 213,—, öS 1490,—. ISBN 3-211-82046-9

Genes expressed in different plant organs and processes are discussed with the emphasis on identifying various regulatory circuits controlling plant gene expression in a temporal and spatial manner. Regulation of foreign genes in transgenic plants is addressed with respect to a number of agriculturally important traits. This book illustrates the complexity of gene expression in plants and outlines strategies towards manipulating specific genes of interest.

MIX
Papier aus verantwortungsvollen Quellen
Paper from responsible sources
FSC® C105338
FSC
www.fsc.org

If you have any concerns about our products,
you can contact us on
ProductSafety@springernature.com

In case Publisher is established outside the EU,
the EU authorized representative is:
Springer Nature Customer Service Center GmbH
Europaplatz 3, 69115 Heidelberg, Germany

Printed by Libri Plureos GmbH
in Hamburg, Germany